Universitext

Kurt Gödel (1906–1978)

S.M. Srivastava

A Course on Mathematical Logic

S.M. Srivastava
Indian Statistical Institute
Kolkata
India
e-mail: smohan@isical.ac.in

ISBN: 978-0-387-76275-3 e-ISBN: 978-0-387-76277-7
DOI: 10.1007/978-0-387-76277-7

Library of Congress Control Number: 2008920049

Mathematics Subject Classification (2000): 03-xx

Printed on acid-free paper

9 8 7 6 5 4 3 2 1

springer.com

To HAIMANTI

Contents

Preface

This book is written on the occassion of the birth centenary year of Kurt Gödel (1906–1978), the most exciting logician of all times, whose discoveries shook the foundations of mathematics. His beautiful technique to examine the whole edifice of mathematics within mathematics itself has been likened, not only figuratively but also in precise technical terms, to the music of Bach and drawings of Escher [4]. It had a deep impact on philosophers and linguists. In a way, it ushered in the era of computers. His idea of arithmetization of formal systems led to the discovery of a universal computer program that simulates all programs. Based on his incompleteness theorems, physicists have propounded theories concerning artifical intelligence and the mind–body problem [10].

The main goal of this book is to state and prove Gödel's completeness and incompleteness theorems in precise mathematical terms. This has enabled us to present a short, distinctive, modern, and motivated introduction to mathematical logic for graduate and advanced undergraduate students of logic, set theory, recursion theory, and computer science. Any mathematician who is interested in knowing what mathematical logic is concerned with and who would like to learn the famous completeness and incompleteness theorems of Gödel should also find this book particularly convenient. The treatment is thoroughly mathematical, and the entire subject has been approached like any other branch of mathematics. Serious efforts have been made to make the book suitable for both instructional and self-reading purposes. The book does not strive to be a comprehensive encyclopedia of logic, nor does it broaden its audience to linguists and philosophers. Still, it gives essentially all the basic concepts and results in mathematical logic.

The main prerequisite for this book is the willingness to work at a reasonable level of mathematical rigor and generality. However, a working knowledge of elementary mathematics, particularly naive set theory and algebra, is required. We suggest [12, pp. 1–15] for the necessary prerequisites in set theory. A good source for the algebra needed to understand some examples and applications would be [7].

Students who want to specialize in foundational subjects should read the entire book, preferably in the order in which it is presented, and work out all the problems. Sometimes we have only sketched the proof and left out the routine arguments for readers to complete. Students of computer science may leave out sections on model theory and arithmetical sets. Mathematicians working in other areas and who want to know about the completeness and incompleteness theorems alone may also omit these sections. However, sections on model theory give applications of logic to mathematics. Chapters 1 to 4, except for Section 2.4 and Sections 5.1 and 5.4, should make a satisfactory course in mathematical logic for undergraduate students.

The book prepares students to branch out in several areas of mathematics related to foundations and computability such as logic, model theory, axiomatic set theory, definability, recursion theory, and computability. Hinman's recent book [3] is the most comprehensive one, with representation in all these areas. Shoenfield's [11] is still a very satisfactory book on logic. For axiomatic set theory, we particularly recommend Kunen [6] and Jech [5]. For model theory, the readers should also see Chang and Keisler [2] and Marker [8]. For recursion theory we suggest [9].

Acknowledgments. I thank M. G. Nadkarni, Franco Parlamento, Ravi A. Rao, B. V. Rao, and H. Sarbadhikari for very carefully reading the entire manuscript and for their numerous suggestions and corrections. Thanks are also due to my colleagues and research fellows at the Stat-Math Unit, Indian Statistical Institute, for their encouragements and help. I fondly acknowledge my daughter Rosy, my son Ravi, and my grandsons Pikku and Chikku for keeping me cheerful while I was writing this book. Last but not least, I shall ever be grateful to my wife, H. Sarbadhikari, for cheerfully putting up with me at home as well at the office all through the period I was working on the book.

1

Syntax of First-Order Logic

The main objects of study of mathematical logic are mathematical theories such as set theory, number theory, and the theory of algebraic structures such as groups, rings, fields, algebraically closed fields, etc., with the aim to develop tools to examine their consistency, completeness, and other similar questions concerning the foundation of these theories. In this chapter we take the first step toward logic and precisely define the notion of a first-order theory.

1.1 First-Order Languages

The objects of study in the natural sciences have physical existence. By contrast, mathematical objects are concepts, e.g., "sets," "belongs to ($\in$)," "natural numbers," "real numbers," "complex numbers," "lines," "curves," "addition," "multiplication," etc.

There have to be initial concepts in a theory. To elaborate it a bit more, note that a concept can be defined in terms of other concepts. For instance, $x - y$ is the unique number z such that $y + z = x$; or if x and y are sets, $x \subset y$ if for every element z, $z \in x$ implies $z \in y$. Thus, "subtraction" can be "defined" in terms of "addition" and "subset ($\subset$)" in terms of "belongs to ($\in$)." At the onset, one begins with a minimal number of undefined concepts. For instance, in set theory, the undefined concepts are "sets" and "belongs to"; in number theory, the undefined concepts are "natural numbers," "zero," and the "successor function"; in the theory of real numbers (seen as an archimedean ordered field), the undefined concepts are "real numbers," "zero," "one," "addition," "multiplication," and "less than." In these examples, we see that there are two groups of concepts: sets or natural numbers or real numbers on the one hand; belongs to, zero, one, successor, addition, multiplication, less than, etc. on the other. Concepts of the first type are the main objects of study; concepts of the second type are used to reflect basic structural properties of the objects of the first type. Then one lists a set of axioms that give the basic structural properties of the objects of study. It is expected that based on these undefined concepts and the axioms, other concepts can be defined. Then the theory is developed by introducing more and more concepts and proving more and more theorems.

Clearly, we ought to have a language to develop a theory. Like any of the natural languages, say Latin, Sanskrit, Tamil, etc., a language suitable for a mathematical theory also has an alphabet. But unlike natural languages, a statement in a mathematical theory is expressed symbolically and has an unambiguous syntactical construction. Before giving precise definitions, we give some examples of statements in some theories that we are familiar with.

Example 1.1.1 Consider the following statement in group theory: For every x there exists y such that $x \cdot y = e$. Here $\cdot$ (dot) is a symbol for the binary group operation and e for the identity element. If we use the symbol $\forall$ to denote "for every" and $\exists$ for "there exists," then we can represent the above statement as follows:

$$\forall x \exists y (x \cdot y = e).$$

Example 1.1.2 The following are two statements in set theory:

$$\forall x \forall y \exists z (x \in z \wedge y \in z),$$

and

$$\neg \exists x \forall y (y \in x).$$

The first statement is a symbolic representation of the statement "Given any two sets x and y, there is a set z that contains both x and y"; the second statement means that "There is no set x that contains all sets y".

We see that the language for a theory should have "variables" to represent the objects of study, e.g., sets in set theory, or elements of a group in group theory, etc., and some logical symbols like $\exists$ (there exists), $\wedge$ (and), $\neg$ (negation), $=$ (equality). These symbols are common to the languages for all theories. We call them logical symbols. On the other hand, there are certain alphabets that represent undefined concepts of a specific theory. For instance, in group theory we use two symbols: the dot $\cdot$ for the group operation and a symbol, say e, for the identity element; in set theory we have a binary relation symbol $\in$ for the undefined concept belongs to.

We make one more observation before giving the first definition in the subject. Mathematicians use many logical connectives and quantifiers such as $\vee$ (or), $\wedge$ (and), $\exists$ (there exists), $\forall$ (for all), $\rightarrow$ (if $\cdots$, then $\cdots$), and $\leftrightarrow$ (if and only if). However, in their reasoning "two statements A and B are both true" if and only if "it is not true that any of A or B is false"; "A implies B" if and only if "either A is false or B is true," etc. This indicates that some of the logical connectives and quantifiers can be defined in terms of others. So, we can start with a few logical connectives and quantifiers. This economy will help in making many proofs quite short.

A first-order language L consists of two types of symbols: **logical symbols** and **nonlogical symbols**. Logical symbols consist of a sequence of **variables** $x_0, x_1, x_2, \ldots$; **logical connectives** $\neg$ (negation) and $\vee$ (disjunction); a **logical quantifier** $\exists$ (existential quantifier) and the **equality symbol** $=$. We call the order in which variables $x_0, x_1, x_2, \ldots$ are listed the **alphabetical order**. These are common to all first-order languages. Depending on the theory, nonlogical symbols of L consist of an (empty or nonempty) set of **constant symbols** $\{c_i : i \in I\}$; for each positive integer n, a set of n-ary **function symbols** $\{f_j : j \in J_n\}$; and a set of n-ary **relation symbols** $\{p_k : k \in K_n\}$.

When it is clear from the context, a first-order language will simply be called a language. Since logical symbols are the same for all languages, to specify a language one has to specify its nonlogical symbols only. To avoid suffixes and for ease in reading, we shall use symbols x, y, z, u, v, w, with or without subscriptss, to denote variables. Any finite sequence of symbols of a language L will be called an **expression** in L.

A language L is called **countable** if it has only countably many nonlogical symbols; it is called **finite** if it has finitely many nonlogical symbols.

Example 1.1.3 The language for set theory has only one nonlogical symbol: a binary relation symbol $\in$ for "belongs to."

Example 1.1.4 The language for group theory has a constant symbol e (for the identity element) and a binary function symbol $\cdot$ (for the group operation).

Example 1.1.5 The language for the theory of rings with identity has two constant symbols 0 and 1 and two binary function symbols $+$ and $\cdot$.

Example 1.1.6 The language for the theory of ordered fields has two constant symbols 0 and 1, two binary function symbols $+$ and $\cdot$, and a binary relation symbol $<$.

A first-order language L' is called an **extension** of another language L if every constant symbol of L is a constant symbol of L' and every n-ary function (relation) symbol of L is an n-ary function (relation) symbol of L'.

Example 1.1.7 The language for the theory of ordered fields is an extension of the language for the theory of rings with identity.

Exercise 1.1.8 Show that the set of all expressions of a countable language is countable.

1.2 Terms of a Language

We now define **terms** of a language L. Broadly speaking, they correspond to algebraic expressions.

The set of all terms of a language L is the smallest set $\mathcal{T}$ of expressions of L that contains all variables and constant symbols and is closed under the following operation: whenever $t_1, \ldots, t_n \in \mathcal{T}$, $f_j t_1 \cdots t_n \in \mathcal{T}$, where f_j is any n-ary function symbol of L. Equivalently, all the terms of a language can be inductively defined as follows: variables and constant symbols are terms of rank 0; if $t_1, \ldots, t_n$ are terms of rank $\leq k$, and if f_j is an n-ary function symbol, then $f_j t_1 \cdots t_n$ is a term of rank at most $k+1$. Thus, the **rank** of a term t is the smallest natural number k such that t is of rank $\leq k$.

Note that the set of variable-free terms is the smallest set $\mathcal{T}'$ of expressions of L that contains all constant symbols and is closed under the following operation: whenever $t_1, \ldots, t_n \in \mathcal{T}'$, then $f_j t_1 \cdots t_n \in \mathcal{T}'$, where f_j is any n-ary function symbol of L.

We shall freely use parentheses and commas in a canonical way for easy readability. For instance, we shall often write $f_j(t_1, \ldots, t_n)$ instead of $f_j t_1 \cdots t_n$, and $t + s$ instead of $+ts$. We shall also drop parentheses when there is no possibility of confusion. Further, we shall adopt the convention of association to the right for omitting parentheses. For instance, instead of writing $t_1 \cdot (t_2 \cdot (t_3 \cdot t_4))$, we shall write $t_1 \cdot t_2 \cdot t_3 \cdot t_4$. It is important to note that the term $((t_1 \cdot t_2) \cdot t_3) \cdot t_4$ is not the same as $t_1 \cdot t_2 \cdot t_3 \cdot t_4$. This term can only be written using parentheses, unless, of course, one writes it as

$$\cdots t_1 t_2 t_3 t_4!$$

Similarly, $(t_1 \cdot t_2) \cdot (t_3 \cdot t_4)$ will stand for

$$\cdot \cdot t_1 t_2 \cdot t_3 t_4!$$

Example 1.2.1 Let L be the language for the theory of rings with identity: L has two constant symbols, 0 and 1, and two binary function symbols, $+$

and $\cdot$. Let $\underline{m}$ denote the term obtained by "adding" 1 to itself m times, i.e., $\underline{m}$ is the term

$$\underbrace{1 + \cdots + 1}_{m \text{ times}};$$

for any term t, let t^n denote the term obtained by "multiplying" t with itself n times, i.e., t^n is the term

$$\underbrace{t \cdot t \ \cdots \ t \cdot t}_{n \text{ times}}.$$

Then $\underline{m}$ and t^n are terms of L. Also any "formal polynomial"

$$\underline{m_0} + \underline{m_1} x + \cdots + \underline{m_n} x^n,$$

x a variable, is a term of L.

Example 1.2.2 Variables are the only terms of the language for set theory because it has no constant and no function symbols.

We define the set of all **subterms** of a term t by induction as follows: t is a subterm of t. If $ft_1 \cdots t_n$, $t_1, \ldots, t_n \in \mathcal{T}$, is a subterm of t, so is each t_i, $1 \leq i \leq n$. An expression is a subterm of t if it is obtained as above. Thus, *the set of all subterms of a term t is the smallest set $\mathcal{S}$* of expressions that contains t and such that whenever $ft_1 \cdots t_n \in \mathcal{S}$, then $t_1, \ldots, t_n \in \mathcal{S}$.

Example 1.2.3 Let t be the term $x \cdot y \cdot z$ of the language for group theory. Then $x \cdot y \cdot z$, x, $y \cdot z$, y, and z are all the subterms of t. Note that $x \cdot y$ is not a subterm of t.

Exercise 1.2.4 List all the subterms of the term

$$x \cdot u + y \cdot v + z \cdot w$$

of the language for the theory of rings.

Let s be a term. We shall write $s[v_1, \ldots, v_n]$ to indicate that variables occurring in s are among $v_1, \ldots, v_n$. If s is a term, $s_{v_1,\ldots,v_n}[t_1, \ldots, t_n]$, or simply $s[t_1, \ldots, t_n]$ when there is no possibility of confusion, denotes the expression obtained from s by simultaneously replacing all occurrences of $v_1, \ldots, v_n$ in s by $t_1, \ldots, t_n$ respectively.

Example 1.2.5 Let s be the term $x \cdot (y + z)$ of the language for the theory of rings with identity. Then

$$s_{x,y,z}[x + z, 1, y \cdot y] = (x + z) \cdot (1 + y \cdot y).$$

Proposition 1.2.6 *Let $s[v_1, \ldots, v_n]$ and let $t_1, \ldots, t_n$ be terms. The expression $s[t_1, \ldots, t_n]$ defined above is a term.*

Proof. We prove the result by induction on the rank of s. If s is a constant symbol c, $s[t_1, \ldots, t_n] = c$; if s is a variable other than the v_i's, then $s[t_1, \ldots, t_n] = s$; if s is v_i for some $1 \leq i \leq n$, $s[t_1, \ldots, t_n] = t_i$. Thus, the assertion is true for terms of rank 0.

Let k be a natural number and assume that the assertion be true for all terms s of rank $\leq k$ (and all variables v_i and all terms t_i). Let $s_j[v_1, \ldots, v_n]$, $1 \leq j \leq m$, be terms of rank $\leq k$, $t_1, \ldots, t_n$ terms, and f an m-ary function symbol. Suppose

$$s[v_1, \ldots, v_n] = f(s_1[v_1, \ldots, v_n], \ldots, s_m[v_1, \ldots, v_n]).$$

Then

$$s[t_1, \ldots, t_n] = f(s_1[t_1, \ldots, t_n], \ldots, s_m[t_1, \ldots, t_n]).$$

By the induction hypothesis, each $s_j[t_1, \ldots, t_n]$ is a term. Hence $s[t_1, \ldots, t_n]$ is a term. The proof is complete by induction on the rank of terms. □

Remark 1.2.7 The above method of proving statements on terms by induction on the rank of terms is a fairly standard one in the subject. Sometimes, in the rest of this book, we may not give the complete argument and just say that the result can be proved by induction on the rank of terms.

1.3 Formulas of a Language

Our next concept is that of an atomic formula of the language L.

An **atomic formula** of a language is defined as follows: if t and s are terms of L, then $t = s$ is an atomic formula of L; if p is an n-ary relation symbol of L and $t_1, \ldots, t_n$ are terms, then $pt_1 \cdots t_n$ is an atomic formula; these are all the atomic formulas of L.

Example 1.3.1 $x \cdot y = 1$, $\underline{i} \cdot (\underline{j} + \underline{k}) = \underline{i} \cdot \underline{j} + \underline{i} \cdot \underline{k}$, $\underline{i} \cdot \underline{i} < \underline{m}$ are atomic formulas of the language for the theory of ordered fields.

Example 1.3.2 $v \in w$, $v = w$, where v, w are variables, are all the atomic formulas of the language for set theory.

A **formula** of a language is inductively defined as follows: every atomic formula is a formula—these are all the formulas of **rank** 0; if A and B are formulas of rank $\leq k$ and v is a variable, then $\neg A$ (the negation of A); $\exists v A$ and $\vee AB$ (the disjunction of A and B) are formulas of rank $\leq k + 1$. The set of strings so obtained are all the formulas of L. Thus, *the set of all formulas of L is the smallest set of all expressions of L that contains all the atomic formulas and that is closed under negation, disjunction, and existential quantification.* Let A be a formula of L. The **rank** of A is the smallest natural number k such that the rank of A is $\leq k$.

From now on, unless otherwise stated, L will denote a first-order language, and by a term (or a formula) we shall mean a term (or a formula) of L.

We shall generally write $A \vee B$ instead of $\vee AB$. In the case of formulas also, we shall use parentheses and commas in a canonical way for easy readability. We adopt the convention of association to the right for omitting parentheses. This means that $A \vee B \vee C$ is to be read as $A \vee (B \vee C)$; $A \vee B \vee C \vee D$ is to be read as $A \vee (B \vee (C \vee D))$; and so on. Note that the formula $(A \vee B) \vee C$ is different from the formula $A \vee (B \vee C)$ and that the parentheses have to be used to write the former formula, unless, of course, one writes it as $\vee \vee ABC$!

Remark 1.3.3 Any term or formula is of the form $Au_1 \cdots u_n$ where A is a symbol and $u_1, \cdots, u_n$ terms or formulas. It should be noted that such a representation of a term or formula is unique. This allows us to define functions or give proofs by induction on the length of terms or formulas.

We now define some other commonly used logical connectives and quantifiers:

$\forall vA$ is an abbreviation of $\neg\exists v\neg A$; $A \wedge B$ abbreviates $\neg(\neg A \vee \neg B)$; $A \rightarrow B$ is an abbreviation of $(\neg A) \vee B$; and $A \leftrightarrow B$ abbreviates $(A \rightarrow B) \wedge (B \rightarrow A)$. Note that according to our convention of omitting parentheses, $A \rightarrow B \rightarrow C$ is to be read as $A \rightarrow (B \rightarrow C)$; $A \rightarrow B \rightarrow C \rightarrow D$ is to be read as $A \rightarrow (B \rightarrow (C \rightarrow D))$; and so on. The connective $\wedge$ is called **conjunction** and the quantifier $\forall$ the **universal quantifier**. Note that we could have added all these symbols into our alphabet. There are several reasons for not doing so. For instance, proof of some results concerning formulas will become long if we do not exercise economy in the number of logical symbols.

A formula of the form $\exists vA$ is called an **instantiation** of A, and a formula of the form $\forall vA$ is called a **generalization** of A. A formula is called **elementary** if it is either an atomic formula or an instantiation of a formula.

Exercise 1.3.4 Show that the set of all formulas is the smallest collection $\mathcal{F}$ of formulas such that each elementary formula is in $\mathcal{F}$ and that is closed under $\neg$ and $\vee$, i.e., whenever $A, B \in \mathcal{F}$, then $\neg A$ and $A \vee B$ are in $\mathcal{F}$ as well.

A **subformula** of a formula A is inductively defined as follows: A is a subformula of itself; if $\neg B$ or $\exists vB$ is a subformula of A, then so is B; if $B \vee C$ is a subformula of A, then B and C are subformulas of A; nothing else is a subformula of A. Thus, *the set of subformulas of A is the smallest set $\mathcal{S}(A)$ of formulas of L that contains A and satisfies the following conditions: whenever $\neg B$ or $\exists vB$ is in $\mathcal{S}(A)$, so is B, and whenever $B \vee C$ is in $\mathcal{S}(A)$, so are B and C.*

Exercise 1.3.5 List all the subformulas of the following formulas:

1. $\forall x \exists y (x \cdot y = e)$.
2. $\forall x \forall y \exists z (x \in z \wedge y \in z)$.
3. $\neg \exists x \forall y (y \in x)$.

(The above formulas should be considered in their unabbreviated forms.)

An occurrence of a variable v in a formula A is **bound** if it occurs in a subformula of the form $\exists v B$; otherwise, the occurrence is called **free**. A variable is said to be free in A if it has a free occurrence in A. We shall write $\varphi[v_0, \ldots, v_n]$ if φ is a formula all of whose free variables belong to the set $\{v_0, \ldots, v_n\}$.

Example 1.3.6 In the formula

$$x \in y \vee \exists x (x \in y),$$

all the occurrences of y are free, the first occurrence of x is free, and other occurrences of x are bound.

A formula with no free variable is called a **closed formula** or a **sentence**. A formula that contains no quantifiers is called an **open formula**.

Exercise 1.3.7 Show that the set of all open formulas is the smallest collection $\mathcal{O}$ of formulas such that each atomic formula is in $\mathcal{O}$ and that is closed under $\neg$ and $\vee$, i.e., whenever $A, B \in \mathcal{O}$, then $\neg A$ and $A \vee B$ are in $\mathcal{O}$ as well.

Let $A[x_0, \ldots, x_{n-1}]$ be a formula whose free variables are among $x_0, \ldots, x_{n-1}$ and x_{n-1} is free in A, where $x_0, \ldots, x_{n-1}$ are the first n variables in alphabetical order. We call

$$\forall x_{n-1} \cdots \forall x_0 A$$

the **closure** of A. Note that if A is closed, it is its own closure.

Let t be a term, v a variable, and A a formula of a language L. We say that **the term t is substitutable for v in A** if for each variable w occurring in t, no subformula of A of the form $\exists w B$ contains an occurrence of v that is free in A.

Example 1.3.8 In the formula

$$x \in y \vee \exists x (x \in y),$$

we can't substitute any term containing x for y.

If t is substitutable for v in A, then $A_v[t]$ designates the expression obtained from A by simultaneously replacing each free occurrence of v in A by t. Similarly, if terms $t_1, \ldots, t_n$ are substitutable in A for $v_1, \ldots, v_n$ respectively, then $A_{v_1, \ldots, v_n}[t_1, \ldots, t_n]$, or $A[t_1, \ldots, t_n]$, when there is no possibility of confusion, called an **instance** of A, will denote the expression obtained from A by simultaneously replacing all free occurrences of $v_1, \ldots, v_n$ in A by $t_1, \ldots, t_n$ respectively. Note that whenever we talk of $A[t_1, \ldots, t_n]$, it will be assumed that $t_1, \ldots, t_n$ are substitutable in A for $v_1, \ldots, v_n$ respectively.

Example 1.3.9 Let A be the formula

$$x \in y \vee \exists x(x \in y).$$

Then $A_x[z]$ is the formula $z \in y \vee \exists x(x \in y)$.

Proposition 1.3.10 *The sequence $A[t_1, \ldots, t_n]$ defined above is a formula.*

Proof. As in the case of the corresponding result on terms, this result is also proved by induction on the rank of formulas. Let $A[v_1, \ldots, v_n]$ be an atomic formula. So A is a formula of the form $p(s_1[v_1, \ldots, v_n], \ldots, s_m[v_1, \ldots, v_n])$, where p is an m-ary predicate symbol and $s_1, \ldots, s_m$ terms of L (p may be the equality symbol. In this case $m = 2$). Then,

$$A[t_1, \ldots, t_n] = p(s_1[t_1, \ldots, t_n], \ldots, s_m[t_1, \ldots, t_n]).$$

By Proposition 1.2.6, $s_j[t_1, \ldots, t_n]$, $1 \leq j \leq m$, are terms. Hence, $A[t_1, \ldots, t_n]$ is a formula. Thus, the assertion is true for formulas of rank 0.

Let k be a natural number and assume that the assertion is true for all formulas of rank $\leq k$ (and all variables v_i and all terms t_i).

Let $B[v_1, \ldots, v_n]$ and $C[v_1, \ldots, v_n]$ be formulas of rank $\leq k$ and let $t_1, \ldots, t_n$ be substitutable for $v_1, \ldots, v_n$ respectively in B and C. If A is the formula $\neg B$, $A[t_1, \ldots, t_n]$ is the expression $\neg B[t_1, \ldots, t_n]$, which is a formula by the induction hypothesis. If A is $B \vee C$, $A[t_1, \ldots, t_n]$ is the expression $B[t_1, \ldots, t_n] \vee C[t_1, \ldots, t_n]$, which is a formula by the induction hypothesis.

Let $B[v, v_1, \ldots, v_n]$ be a formula of rank k and let v be distinct from the v_i's. Suppose A is the formula $\exists v B$. Then $A[t_1, \ldots, t_n]$ is the expression $\exists v B[v, t_1, \ldots, t_n]$. This is clearly a formula by the induction hypothesis. Thus the assertion is true for all formulas of rank $k + 1$. Our proof is complete by induction on the rank of formulas. □

Remark 1.3.11 The above method of proving results by induction on the rank of formulas is a fairly standard method in the subject. Sometimes, in the rest of this book, we may not give the complete argument and just say that the result can be proved by induction on the rank of formulas.

So far, we have been describing the "syntax," i.e., rules for arranging symbols into terms and sentences, of a theory. Here a sentence is just a string of symbols from the language of the theory (without having a meaning). One may consider this to be a useless representation of a sentence.

But it is far from so. Logical connectives and quantifiers have an intended logical meaning, so that whatever A may be, $\neg A \vee A$ is "true"; for any term t, $t = t$ is "true"; $A \vee B$ is "true" if and only if at least one of A and B is "true"; and so on. Thus, quite often, the structure of a formula itself helps us to make inferences about the formula. We are now in a very good situation: we have a precise definition of a sentence; it exists concretely, as the string of symbols that we see; and we can make some inferences about it from its syntactical structure. Of course, we should know what is inference and how it is done. This will be specified later in this book.

1.4 First-Order Theories

A **first-order theory** or simply a **theory** T consists of a first-order language L and a set of formulas of L. These formulas are called **nonlogical axioms** of T. By terms or formulas of T, we shall mean terms or formulas respectively of the language for T. The language for T will also be denoted by $L(T)$. A theory is called **countable** if its language is countable. It is **finite** if the set of all nonlogical symbols is finite. In general, a theory T whose set of all nonlogical symbols is of cardinality κ, κ an infinite cardinal, is called a **κ-theory**.

Example 1.4.1 **Group theory** is the theory whose nonlogical symbols are a constant symbol e and a binary function symbol $\cdot$ and whose nonlogical axioms are the following formulas: (Below, x, y, and z denote the first three variables.)

1. $\forall x \forall y \forall z (x \cdot (y \cdot z) = (x \cdot y) \cdot z)$.
2. $\forall x (x \cdot e = x \wedge e \cdot x = x)$.
3. $\forall x \exists y (x \cdot y = e \wedge y \cdot x = e)$.

Example 1.4.2 **The theory of abelian groups** is the theory whose nonlogical symbols are a constant symbol 0 and a binary function symbol $+$ and whose nonlogical axioms are the following formulas:

1. $\forall x \forall y \forall z (x + (y + z) = (x + y) + z)$.
2. $\forall x (x + 0 = x \wedge 0 + x = x)$.
3. $\forall x \exists y (x + y = 0 \wedge y + x = 0)$.
4. $\forall x \forall y (x + y = y + x)$.

Example 1.4.3 The language for **the theory of rings with identity** has two constant symbols, 0 and 1, and two binary function symbols, $+$ and $\cdot$. The nonlogical axioms of this theory are the axioms (1)–(4) of abelian groups together with the following axioms:

5 $\forall x \forall y \forall z (x \cdot (y \cdot z) = (x \cdot y) \cdot z)$.
6 $\forall x (x \cdot 1 = x \wedge 1 \cdot x = x)$.

7 $\forall x \forall y \forall z (x \cdot (y + z) = x \cdot y + x \cdot z)$.
8 $\forall x \forall y \forall z ((y + z) \cdot x = y \cdot x + z \cdot x)$.

Example 1.4.4 **The theory of fields** has the same language as that of the theory of rings with identity; its nonlogical axioms are the axioms (1)–(8) of the the theory of rings with identity together with the following axioms:

9 $\forall x \forall y (x \cdot y = y \cdot x)$.
10 $\forall x (\neg (x = 0) \rightarrow \exists y (x \cdot y = 1 \wedge y \cdot x = 1))$.

Example 1.4.5 Let L be a language with only one nonlogical symbol—a binary relation symbol $<$. The theory LO (**the theory of linearly ordered sets**) is the theory whose language is L and whose nonlogical axioms are the following:

1 $\forall x \neg (x < x)$.
2 $\forall x \forall y \forall z ((x < y \wedge y < z) \rightarrow x < z)$.
3 $\forall x \forall y (x < y \vee x = y \vee y < x)$.

Example 1.4.6 **The theory of dense linearly ordered sets**, denoted by DLO, is obtained from LO by adding the following axioms:

4 $\forall x \forall y ((x < y) \rightarrow \exists z (x < z \wedge z < y))$.
5 $\forall x \exists y (y < x)$.
6 $\forall x \exists y (x < y)$.

Exercise 1.4.7 Express the axioms of an equivance relation as formulas of a suitable first-order language.

Example 1.4.8 Let F be the theory of fields. For each $m \geq 1$, let A_m be the formula $\neg (\underline{m} = 0)$. The theory obtained by adding each A_m to the set of axioms of F as an axiom is called **the theory of fields of characteristic** 0.

Example 1.4.9 Let F be the theory of fields. Let L be an extension of the language for the theory of rings with identity obtained by adding a new binary predicate symbol $<$. Consider the theory OF whose language is L and whose nonlogical axioms are all the nonlogical axioms of F and the following axioms:

11 $\forall x \neg (x < x)$.
12 $\forall x \forall y \forall z ((x < y \wedge y < z) \rightarrow x < z)$.
13 $\forall x \forall y (x < y \vee x = y \vee y < x)$.
14 $\forall x \forall y (\neg (x < y \vee x = y) \rightarrow y < x)$.
15 $\forall x \forall y (x < y \rightarrow \forall z (x + z < y + z))$.
16 $\forall x \forall y ((0 < x \wedge 0 < y) \rightarrow 0 < x \cdot y)$.

The theory OF is known as the **theory of ordered fields**.

Example 1.4.10 We now give some axioms of number theory, which plays an important role in logic. We designate this theory by N. The nonlogical symbols of N are a constant symbol 0, a unary function symbol S (which designates the successor function), two binary function symbols $+$ and $\cdot$, and a binary relation symbol $<$. The nonlogical axioms of N are:

1. $\forall x(\neg(Sx = 0))$.
2. $\forall x \forall y(Sx = Sy \rightarrow x = y)$.
3. $\forall x(x + 0 = x)$.
4. $\forall x \forall y(x + Sy = S(x + y))$.
5. $\forall x(x \cdot 0 = 0)$.
6. $\forall x \forall y(x \cdot Sy = (x \cdot y) + x)$.
7. $\forall x(\neg(x < 0))$.
8. $\forall x \forall y(x < Sy \leftrightarrow (x < y \vee x = y))$.
9. $\forall x(\forall y(x < y \vee x = y \vee y < x)$.

For any nonnegative integer n, the term

$$\underbrace{S \cdots S}_{m \text{ times}} 0$$

will be denoted by k_n. These terms are called **numerals**. Note that k_0 is the constant symbol 0.

Example 1.4.11 Peano arithmetic is the theory obtained from N by deleting the last axiom and adding the following axiom schema, called the **induction axiom schema**: for every formula $A[v]$, the formula

$$A_v[0] \rightarrow \forall v(A \rightarrow A_v[Sv]) \rightarrow A$$

is called an induction axiom. This theory will be denoted by PA.

Example 1.4.12 We give below the axioms of set theory. This theory is called **Zermelo–Fraenkel set theory**, and is designated by ZF. To convey the content of the axioms better, we shall state the axioms informally in words also:

1. **Set Existence.** *There exists a set.* This is expressed by the following formula:
$$\exists x(x = x).$$
2. **Extensionality.** *Two sets are the same if they have exactly same members:*
$$\forall x \forall y(\forall z(z \in x \leftrightarrow z \in y) \rightarrow x = y).$$
3. **Comprehension (subset) schema.** For each formula $\varphi[x, w_1, \ldots, w_n]$, the following is an axiom.

$$\forall z \forall w_1 \cdots \forall w_n(\exists y \forall x(x \in y \leftrightarrow x \in z \wedge \varphi)).$$

This axiom says that *given any "property of sets" expressed by a formula* $\varphi[x, w_1, \ldots, w_n]$, *for any fixed parameters* $w_1, \ldots, w_n$ *and for any set* z, *there is a set* y *that consists precisely of those* $x \in z$ *that satisfy* $\varphi[x, w_1, \ldots, w_n]$.
By extensionality, it can be proved that such a set y is unique. It is assumed that the variables x, y, z and the w_i's are distinct.

4. **Replacement schema.** For every formula $\varphi[x, y, z, u_1, \ldots, u_n]$, the following formula is an axiom:

$$\forall z \forall u_1 \cdots \forall u_n (\forall x \in z \exists! y \varphi \to \exists v \forall x (x \in z \to \exists y (y \in v \land \varphi))),$$

where $\exists! y \varphi$ abbreviates the formula

$$\varphi \land \forall u (\varphi_y[u] \to u = y).$$

This axiom together with comprehension says that the range of a "function" on a set z that is defined by a formula φ is a set.

5. **Pairing.** *Given sets* x *and* y, *there is a set* z *that contains both* x *and* y*:*

$$\forall x \forall y \exists z (x \in z \land y \in z).$$

This axiom together with comprehension helps us to talk of sets of the form $\{x\}, \{x, y\}, \{x, y, z\}$, etc.

6. **Union.** *Given any set* x, *there is a set* y *that contains all those* z *that belong to a member of* x*:*

$$\forall x \exists y \forall z \forall u (u \in x \land z \in u \to z \in y).$$

This axiom together with comprehension will imply that the union of a family of sets is a set.

7. **Power Set.** *Given any set* x, *there is a set* y *that contains all subsets* z *of* x*:*

$$\forall x \exists y \forall z (\forall u (u \in z \to u \in x) \to z \in y).$$

This axiom together with comprehension will enable us to define the power set of a set.

8. **Infinity.** Based on the axioms introduced so far, it can be "proved" that the empty set exists, which we shall denote by 0. The following formula is an axiom:

$$\exists x (0 \in x \land \forall y (y \in x \to y \cup \{y\} \in x)).$$

Without the infinity axiom, we can't prove the existence of an "infinite" set; without this axiom, we can't prove that there is a set containing all natural numbers.

9. **Foundation.** This is the most unintuitive axiom. It is the following formula:

$$\forall x(\exists y(y \in x) \rightarrow \exists y(y \in x \wedge \neg\exists z(z \in x \wedge z \in y))).$$

It says that *the binary relation* $\in$ *is well-founded on every nonempty set.* It rules out the existence of a set that contains itself. It also has the effect of restricting the domain of discourse of set theory to those sets where mathematics actually takes place.

2

Semantics of First-Order Languages

In the previous chapter, we presented the syntactical notions pertaining to first-order theories. However, in general, mathematical theories are not developed syntactically. There is, of course, one serious exception to this: essentially, due to its foundational nature, axiomatic set theory is developed syntactically. Since set theory is needed for proving independence results, the syntactical approach is quite important for mathematics. In this chapter we give the semantics of first-order languages to connect the syntactical description of a theory with the setting in which a mathematical theory is generally developed.

Recall that instead of beginning with the syntactical object group theory, in practice, one begins by defining a group as a nonempty set G with a specified element e and a binary operation $\cdot : G \times G \to G$ satisfying the following three conditions:

1. For every a, b, c in G,
$$a \cdot (b \cdot c) = (a \cdot b) \cdot c.$$
2. For every $a \in G$,
$$a \cdot e = e \cdot a = a.$$
3. For every $a \in G$, there is a $b \in G$ such that
$$a \cdot b = b \cdot a = e.$$

Thus a group consists of a nonempty set G with "interpretations" or "meanings" of the nonlogical symbols $\cdot$ (a binary function symbol) and e (a constant symbol) such that all the axioms of group theory are "satisfied." Further, a statement of the language for group theory is called a theorem if it is satisfied in all groups. Thus, to give the connection we are looking for, first we should define the interpretation or a structure of a language L as a nonempty set A together with the interpretations or meanings of all the nonlogical symbols of L. This is known as the semantics of L. Then models of a theory T are those structures of the language for T in which all nonlogical axioms are true.

In this chapter we shall define the structure of a language, truth or satisfiability in a structure, and the model of a theory. Finally, we shall generalize the notion of homomorphisms, isomorphisms, subgroups, subfields, etc. in the general setting of models of a theory.

2.1 Structures of First-Order Languages

A **structure** or an **interpretation** of a first-order language L consists of (i) a nonempty set M (called the **universe** of the structure), (ii) for each constant symbol c of L, a fixed element $c_M \in M$, (iii) for each n-ary function symbol f of L, an n-ary map $f_M : M^n \to M$, and (iv) for each n-ary relation symbol p of L, an n-ary relation $p_M \subset M^n$ on M. The interpretation of $=$ is always taken to be the equality relation in M.

The elements of the universe M are called **individuals of the structure**; the f_M's the **individual functions of the structure**; and the p_M's, the **individual predicates**; c_M, f_M, and p_M will be called interpretations of the constant symbol c, of the function symbol f, and of the predicate symbol p respectively.

Any group is a structure of the language for group theory; the usual set of real numbers with the usual 0, 1, $+$, $\cdot$, and $<$ is a structure for the language of the theory of ordered fields. Note that which statement of L is true in the structure and which is not, is not relevant in the notion of a structure of L. For instance, the set of all natural numbers $\mathbb{N} = \{0, 1, 2, \ldots\}$ as the universe, 0 as the interpretation of e, and $+$ the interpretation of $\cdot$ is a structure of the language for group theory even though it is not a group.

Example 2.1.1 Let $\mathbb{N}$ be the set of all natural numbers and 0, 1, $+$, $\cdot$, and $<$ have the usual meanings. Further, let $S(n) = n + 1$, $n \in \mathbb{N}$. This is a structure for the language of the theory N defined in Chapter 1. This structure will be called the **standard structure of** N.

Let L be an extension of L' and M a structure of L. By ignoring the interpretations of those nonlogical symbols of L that are not symbols of L', we get a structure M' of L'. We call M' the **restriction** of M to L' and denote it by $M|L'$. In this case we shall also call M an **expansion** of M' to L.

Recall that all variable-free terms can be obtained starting from constant symbols and iterating function symbols on them. So, we can define the interpretation or meaning t_M of each variable-free term t of L in M by induction on the rank of t. The interpretation of a constant symbol c is already given by the structure, namely c_M. If $t_1, \ldots, t_n$ are variable-free terms whose interpretations have been defined and if f is an n-ary function symbol of L, then we define

$$(ft_1 \cdots t_n)_M = f_M((t_1)_M, \ldots, (t_n)_M).$$

By induction on the rank of terms, it is easy to see that we have defined t_M for each variable-free term t of L.

Example 2.1.2 Let L be the language for the theory of rings with identity. For each positive integer m, let $\underline{m}$ denote the term obtained by "adding" 1 to itself m times. Let $P(x)$ be a polynomial expression whose coefficients are of the form $\underline{m}$, i.e., $P(x)$ is a term of the form

$$\underline{m_0} + \underline{m_1}x + \cdots + \underline{m_n}x^n,$$

where x is a variable. Let R be a ring with identity. Then the interpretation of $\underline{m}$ in R is the element $m \in R$ obtained by adding the multiplicative identity of R to itself m times, and for any variable-free term t, the interpretation of $P_x[t]$ in R is the element $P(t_M)$ of R.

2.2 Truth in a Structure

In this section, we shall define when a formula of L is true and when it is false in a structure of L. Note that if we have a structure of L with universe M and we would like to know whether there is an element $a \in M$ satisfying a formula $\varphi[x]$, we have a bit of a problem because φ is a syntactical object, and elements of M are not. To circumvent this problem, given a structure of L with universe M, we first describe an extension L_M of the language L.

Given L and and a structure of L with universe M, let L_M be the first-order language obtained from L by adding a new constant symbol i_a for each $a \in M$. The symbol i_a is called the **name** of a. We regard M itself as the expansion of M to L_M by setting the interpretation of i_a to be a for each $a \in M$.

We are now in a position to define when a formula of L is true or valid or satisfiable in the structure M. To achieve this, we define the notion of truth of a closed formula or a sentence of L_M in the structure M. The definition is based on the well-known intended meanings of logical connectives $\vee$ and $\neg$

and that of the existential quantifier $\exists$. The notion of truth will be defined by defining a function from the set of all closed formulas of L_M to the set {true, false} satisfying some conditions. This will be done by induction on the rank of sentences of L_M. If a sentence takes the value true, we shall say that the sentence is true or valid in M; otherwise, it is said to be false in M.

Recall that formulas have been defined inductively starting from atomic formulas and iterating $\neg$, $\vee$, and $\exists$ on them. A variable-free atomic formula is of the form $pt_1 \cdots t_n$, where p is an n-ary relation symbol (including $=$) and $t_1, \ldots, t_n$ are variable-free terms. We say that $pt_1 \cdots t_n$ is true in the structure if

$$p_M((t_1)_M, \ldots, (t_n)_M)$$

holds, i.e.,

$$((t_1)_M, \ldots, (t_n)_M) \in p_M \subset M^n.$$

Otherwise, we say that $pt_1 \cdots t_n$ is false in the structure. A sentence $\neg A$ is true if and only if A is false. A sentence $A \vee B$ is true if either A is true or B is true. Finally, a sentence $\exists v A$ is true if $A_v[i_a]$ is true for some $a \in M$. We say that a formula A of L_M is true in the structure if its closure is true in the structure. If a formula A of L is true in a structure M of L, we also say that A is **valid in the structure** and write $M \models A$. If A is not valid in M, we write $M \not\models A$.

Note that if A and B are closed formulas, then

$$M \not\models A \Longleftrightarrow M \models \neg A$$

and

$$M \models A \vee B \Longleftrightarrow M \models A \text{ or } M \models B.$$

Exercise 2.2.1 Give an example of a formula (necessarily not closed) of the language for the theory N that is neither true nor false in the standard structure $\mathbb{N}$ of N. Similarly, give examples of formulas A and B of the language for the theory N such that $A \vee B$ is valid in the standard structure $\mathbb{N}$ but neither A nor B is valid in $\mathbb{N}$.

Exercise 2.2.2 Show the following:

1. A sentence $A \wedge B$ is valid in a structure if and only if both A and B are valid in the structure.
2. A sentence of the form $\forall v \varphi[v]$ is valid in a structure with universe M if and only if for each $a \in M$, the sentence $\varphi_v[i_a]$ of L_M is valid in the structure.
3. A sentence of the form $A \to B$ is valid in a structure if and only if either A is false or B true in the structure.
4. A sentence of the form $A \leftrightarrow B$ is valid in a structure if and only if either both A and B are valid or both are not valid in the structure.

Exercise 2.2.3 Let $A[v_1, \ldots, v_n]$ be a formula and $t_1, \ldots, t_n$ variable-free terms of L. Show that the formulas

$$\forall v_1 \cdots \forall v_n A \rightarrow A[t_1, \ldots, t_n]$$

and

$$A[t_1, \ldots, t_n] \rightarrow \exists v_1 \cdots \exists v_n A$$

are valid in all structures of L.

2.3 Model of a Theory

A **model** of a first-order theory T is a structure of $L(T)$ with universe M in which all nonlogical axioms of T are valid. For instance, any group is a model of group theory. On the other hand, the set $\mathbb{N}$ of natural numbers together with the usual 0 and $+$ as the interpretations for e and $\cdot$ respectively is definitely a structure for the language of group theory but not a model of group theory.

A formula A of T that is true in all models of T is called **valid** in T. One writes $T \models A$ if A is valid in T. If A is not valid in some model of T, we shall write $T \not\models A$

Example 2.3.1 Let L be a first-order language with no nonlogical symbol. For each $n > 1$, let A_n be the formula

$$\exists x_0 \cdots \exists x_{n-1} \wedge_{0 \le i < j < n} \neg(x_i = x_j).$$

Suppose T is the theory whose language is L and whose axioms are $A_2, A_3, \ldots$. Then models of T are precisely the infinite sets.

Exercise 2.3.2 A field $\mathbb{K}$ is called algebraically closed if every nonconstant polynomial over $\mathbb{K}$ has a root in $\mathbb{K}$. The **characteristic of a field** $\mathbb{K}$ is the least integer $n \ge 2$ such that

$$\underbrace{1 + \cdots + 1}_{n \text{ times}} = 0.$$

If such an integer n exists, it must be prime. If $\mathbb{K}$ is not of characteristic n for any $n > 1$, $\mathbb{K}$ is said to be of **characteristic** 0.

(i) Show that every algebraically closed field is infinite.
(ii) Define a first-order theory ACF whose models are precisely the algebraically closed fields.
(iii) Define a first-order theory $ACF(p)$, $p \ge 2$ a prime, whose models are precisely the algebraically closed fields of characteristic p.

(iv) Define a first-order theory $ACF(0)$ whose models are precisely the algebraically closed fields of characteristic 0.

Example 2.3.3 Show that the set of all natural numbers

$$\mathbb{N} = \{0, 1, \ldots\}$$

with the usual meanings of S (the successor function), $+$, $\cdot$, and $<$ is a model of the theory N and also of Peano arithmetic. This model will be called the **standard model** of N or of Peano arithmetic.

Exercise 2.3.4 Let L be an extension of L', M a structure of L, and M' the restriction of M to L'. Note that M and M' have the same individuals. Use the same constant as a name for an individual in M and M'. Show that a statement of $L'_{M'}$ is valid in M' if and only if it is valid in M.

In mathematical parlance, valid statements in T are called theorems. So, in order to decide whether a statement is a theorem of T, one has to show that it is true in all models of T. But a sentence is a well-formed finite sequence of symbols. So it is not unreasonable to expect to give a finitary and also constructive notion of theorem. This is where the famous program of the great German mathematician David Hilbert enters. Hilbert proposed to write down a set of axioms and a set of rules of inference to infer a formula from its syntactical structure, and call a sentence a theorem if it can be inferred from axioms by using certain logical rules of inference. We shall elaborate on Hilbert's program in the next few chapters.

2.4 Embeddings and Isomorphisms

In this section we introduce notions analogous to subgroups of a group, isomorphisms of rings, isomorphic fields, etc. in the general context of first-order logic.

In the rest of this section, unless otherwise stated, M and N will denote structures of a fixed first-order language L.

For the sake of brevity, a sequence $(a_1, \ldots, a_n) \in N^n$ will sometimes be denoted by $\overline{a}$ and $(i_{a_1}, \ldots, i_{a_n})$ by $i_{\overline{a}}$. Further, for any map $\alpha : N \to M$, $\alpha(\overline{a})$ will stand for the sequence $(\alpha(a_1), \ldots, \alpha(a_n))$.

An **embedding** of N into M is a one-to-one map $\alpha : N \to M$ satisfying the following conditions:

(1) For every constant symbol c of L,

$$\alpha(c_N) = c_M.$$

(2) For every n-ary function symbol f of L and every $\overline{a} \in N^n$,

$$\alpha(f_N(\overline{a})) = f_M(\alpha(\overline{a})).$$

(2) For every n-ary relation symbol p of L and every $\overline{a} \in N^n$,

$$p_N(\overline{a}) \Longleftrightarrow p_M(\alpha(\overline{a})),$$

i.e.,

$$\overline{a} \in P_N \Longleftrightarrow \alpha(\overline{a}) \in P_M.$$

If, moreover, α is onto M, we call $\alpha : N \to M$ an **isomorphism**. In this case, M and N are called **isomorphic structures**.

If $N \subset M$ and the inclusion map $N \hookrightarrow M$ is an embedding, then N is called a **substructure** of M.

Remark 2.4.1 Let N be a subset of a structure M such that for each constant symbol c, $c_M \in N$, and for every function symbol f, N is closed under f_M. We then make N a substructure of M by setting

(i) for every constant symbol c of L,

$$c_N = c_M,$$

(ii) for every n-ary relation symbol p,

$$p_N = p_M \cap N^n,$$

the restriction of p_M to N, and

(iii) for every n-ary function symbol f,

$$f_N = f_M | N^n,$$

the restriction of f_M to N^n.

Example 2.4.2 Let L be the language for group theory. If H is a subgroup of a group G, then H is a substructure of G. If G and H are groups, then a group isomorphism $\alpha : G \to H$ is an isomorphism from the structure G to the structure H.

Exercise 2.4.3 Show that any two countable models $\mathbb{Q}_1$ and $\mathbb{Q}_2$ of DLO are isomorphic.

Hint: Let $\{r_n\}$ and $\{s_m\}$ be enumerations of $\mathbb{Q}_1$ and $\mathbb{Q}_2$ respectively. Set $n_0 = 0$ and $m_0 = 0$. Suppose for some i, $n_0, \ldots, n_{2i}$ and $m_0, \ldots, m_{2i}$ have been defined so that the the map f defined by

$$f(r_{n_j}) = s_{m_j},\ 0 \leq j \leq 2i,$$

is injective and order-preserving. Now let m_{2i+1} be the first natural number k such that s_k is different from each of s_{m_j}, $j \leq 2i$. Show that there is a natural number l such that r_l is different from each of r_{n_j}, $j \leq 2i$, and the extension of f sending r_l to $s_{m_{2i+1}}$ is order-preserving. Set n_{2i+1} to be the first such l. Thus, the map $f(r_{n_j}) = s_{m_j}$, $j \leq 2i+1$, is injective and order-preserving. Now define n_{2i+2} to be the first natural number l such that r_l is different from each of r_{n_j}, $j \leq 2i+1$. Again observe that there is a natural number k such that s_k is different from each of s_{m_j}, $j \leq 2i+1$, and the extension of the above map by defining $f(r_{n_{2i+2}}) = s_k$ is order-preserving. Set s_{2i+2} to be the least such k.

Proposition 2.4.4 *Let $\alpha : N \to M$ be an embedding, $t[v_1, \ldots, v_n]$ a term of L, and $\overline{a} \in N^n$. Then*

$$\alpha(t[i_{\overline{a}}]_N) = t[i_{\alpha(\overline{a})}]_M.$$

Proof. We prove the result by induction on the rank of t. If t is a variable v_i, then both the terms equal $\alpha(a_i)$. If t is a constant c, then the term on the left is $\alpha(c_N)$ and that on the right is c_M. They are equal because α is an embedding.

Now assume that the result is true for $t_1, \ldots, t_k$ and t is the term $f(t_1, \ldots, t_k)$. Then

$$\begin{aligned}\alpha(t[i_{\overline{a}}]_N) &= \alpha(f_N(t_1[i_{\overline{a}}]_N, \ldots, t_k[i_{\overline{a}}]_N)) \\ &= f_M(\alpha(t_1[i_{\overline{a}}]_N), \ldots, \alpha(t_k[i_{\overline{a}}]_N)) \\ &= f_M(t_1[i_{\alpha(\overline{a})}]_M, \ldots, t_k[i_{\alpha(\overline{a})}]_M) \\ &= t[i_{\alpha(\overline{a})}]_M.\end{aligned}$$

The first equality holds by the definition of $t[i_{\overline{a}}]_N$, the second equality holds because α is an embedding, the third equality holds by the induction hypothesis, and the fourth equality holds by the definition of $t[i_{\alpha(\overline{a})}]_M$.

The proof is complete. □

Proposition 2.4.5 *Let $\alpha : N \to M$ be an embedding, $\varphi[v_1, \ldots, v_n]$ an open formula of L, and $\overline{a} \in N^n$. Then*

$$N \models \varphi[i_{\overline{a}}] \Longleftrightarrow M \models \varphi[i_{\alpha(\overline{a})}]. \qquad (*)$$

Proof. Recall that the set of all open formulas is the smallest class of formulas that contains all atomic formulas and is closed under $\neg$ and $\vee$. So, the result will be proved if we show that the set of formulas φ satisfying $(*)$ contains all atomic formulas and is closed under $\neg$ and $\vee$.

By the definition of the truth in a structure, and the definition of embedding, $(*)$ holds for formulas of the form $t = s$, as well as, for atomic formulas of the form $p(t_1, \ldots, t_n)$.

Now assume that φ is the formula $\neg\psi$ and the result is true for ψ. Then

$$\begin{aligned} N \models \varphi[i_{\overline{a}}] &\iff N \not\models \psi[i_{\overline{a}}] \\ &\iff M \not\models \psi[i_{\alpha(\overline{a})}] \\ &\iff M \models \varphi[i_{\alpha(\overline{a})}]. \end{aligned}$$

The first and the last equivalences hold because the formulas $\psi[i_{\overline{a}}]$ and $\psi[i_{\alpha(\overline{a})}]$ are closed; the second equivalence holds by the induction hypothesis.

The case φ of the form $\psi \vee \eta$ is dealt with similarly:

$$\begin{aligned} N \models \varphi[i_{\overline{a}}] &\iff N \models \psi[i_{\overline{a}}] \text{ or } N \models \eta[i_{\overline{a}}] \\ &\iff M \models \psi[i_{\alpha(\overline{a})}] \text{ or } M \models \eta[i_{\alpha(\overline{a})}] \\ &\iff M \models \varphi[i_{\alpha(\overline{a})}]. \end{aligned}$$

The proof is complete. □

Exercise 2.4.6 Let $\alpha : N \to M$ be a map such that for every open formula $\varphi[v_1, \ldots, v_n]$ of L and every $\overline{a} \in N^n$,

$$N \models \varphi[i_{\overline{a}}] \iff M \models \varphi[i_{\alpha(\overline{a})}].$$

Show that φ is an embedding.

Hint: To show that for any constant symbol c, $\alpha(c_N) = c_M$, let the formula $\varphi[x]$ be $c = x$ and consider $\varphi[i_{c_N}]$; to show that for $a, b \in N$, $\alpha(a) = \alpha(b)$ implies $a = b$, let $\varphi[x, y]$ be the formula $x = y$ and consider $\varphi[i_a, i_b]$.

Theorem 2.4.7 *Let $\alpha : N \to M$ be an isomorphism and $\varphi[v_1, \ldots, v_n]$ a formula of L_N. Then for every $\overline{a} \in N^n$,*

$$N \models \varphi[i_{\overline{a}}] \iff M \models \varphi[i_{\alpha(\overline{a})}]. \qquad (**)$$

In particular, for every sentence φ of L, $N \models \varphi$ if and only if $M \models \varphi$.

Proof. Since an isomorphism is an embedding, by the arguments contained in the proof of Proposition 2.4.5, the set of all formulas φ satisfying (**) contains all atomic formulas and is closed under $\neg$ and $\vee$.

Let $\varphi[v_1, \ldots, v_n]$ be a formula of the form $\exists v \psi$, v different from each of the v_i. Suppose (**) holds for ψ and all $(a, a_1, \ldots, a_n) \in N^{n+1}$. To complete the proof, we now have only to show that (**) holds for φ and every $\overline{a} \in N^n$. So, we take any $\overline{a} \in N^n$. Then,

$$\begin{aligned} N \models \varphi[i_{\overline{a}}] &\iff N \models \psi[i_a, i_{\overline{a}}] \text{ for some } a \in N \\ &\iff M \models \psi[i_{\alpha(a)}, i_{\alpha(\overline{a})}] \text{ for some } a \in N \\ &\iff M \models \psi[i_b, i_{\alpha(\overline{a})}] \text{ for some } b \in M \\ &\iff M \models \varphi[i_{\alpha(\overline{a})}]. \end{aligned}$$

The first equivalence holds by the definition of validity in N, the second equivalence holds by the induction hypothesis, the third equivalence holds because α is surjective, and the last equivalence holds by the definition of validity in M.

The proof is complete □

An embedding $\alpha : N \to M$ is called an **elementary embedding** if for every formula $\varphi[v_1, \ldots, v_n]$ and every $\overline{a} \in N^n$,

$$N \models \varphi[i_{\overline{a}}] \Longleftrightarrow M \models \varphi[i_{\alpha(\overline{a})}].$$

If $N \subset M$ and the inclusion $N \hookrightarrow M$ is an elementary embedding, then we say that N is an **elementary substructure** of M or that M is an **elementary extension** of N. The structures N and M are called **elementarily equivalent** if for every closed formula φ,

$$N \models \varphi \Longleftrightarrow M \models \varphi.$$

We write $N \equiv M$ if N and M are elementarily equivalent. Clearly, $\equiv$ is an equivalence relation on the class of all structures of L.

Remark 2.4.8 By Theorem 2.4.7, two structures N and M are elementarily equivalent if they are isomorphic. Later in this book we shall show that any two algebraically closed fields of characteristic 0 are elementarily equivalent. But there exist two algebraically closed fields $\mathbb{F}_1$ and $\mathbb{F}_2$ of characteristic 0 such that $|\mathbb{F}_1| \neq |\mathbb{F}_2|$. Hence, elementarily equivalent structures need not be isomorphic.

Theorem 2.4.9 *Let N be a substructure of M. Then N is an elementary substructure of M if and only if for every formula $\varphi[v, v_1, \ldots, v_n]$ and for every $\overline{a} \in N^n$, if there is $b \in M$ satisfying*

$$M \models \varphi[i_b, i_{\overline{a}}],$$

then there is $b \in N$ satisfying

$$M \models \varphi[i_b, i_{\overline{a}}].$$

Proof. Let N be an elementary substructure of M. Take a formula $\varphi[v, v_1, \ldots, v_n]$. Let $\overline{a} \in N^n$ and suppose there is $b \in M$ satisfying

$$M \models \varphi[i_b, i_{\overline{a}}].$$

This means that

$$M \models \exists v \varphi[v, i_{\overline{a}}].$$

Since N is an elementary substructure of M, we have

$$N \models \exists v \varphi[v, i_{\overline{a}}].$$

So, there is $b \in N$ satsfying

$$N \models \varphi[i_b, i_{\overline{a}}].$$

Since N is an elementary substructure of M, it follows that

$$M \models \varphi[i_b, i_{\overline{a}}].$$

We prove the if part of the result by showing that for every formula $\psi[v_1, \ldots, v_n]$ and for every $\overline{a} \in N^n$,

$$M \models \psi[i_{\overline{a}}] \Longleftrightarrow N \models \psi[i_{\overline{a}}]. \qquad (*)$$

We shall prove $(*)$ by induction on the rank of ψ. By Proposition 2.4.5, $(*)$ is true for all atomic formulas. Arguing as in the proof of that proposition, we can show that if $(*)$ is true for φ, it is true for $\neg\varphi$, and if φ and ψ satisfy $(*)$, so does $\varphi \vee \psi$.

Now assume that $\varphi[v_1, \ldots, v_n]$ is a formula of the form $\exists v\psi[v, v_1, \ldots, v_n]$ and $(*)$ holds for ψ and every $(a, a_1, \ldots, a_n) \in N^{n+1}$. Take $\overline{a} \in N^n$.

Suppose

$$N \models \varphi[i_{\overline{a}}].$$

Then there is $b \in N$ such that

$$N \models \psi[i_b, i_{\overline{a}}].$$

By the induction hypothesis,

$$M \models \psi[i_b, i_{\overline{a}}].$$

So,

$$M \models \varphi[i_{\overline{a}}].$$

Now assume that

$$M \models \varphi[i_{\overline{a}}].$$

So there is $b \in M$ such that

$$M \models \psi[i_b, i_{\overline{a}}].$$

By our assumptions, there is $b \in N$ such that

$$M \models \psi[i_b, i_{\overline{a}}].$$

By the induction hypothesis,

$$N \models \psi[i_b, i_{\overline{a}}].$$

Thus,

$$N \models \varphi[i_{\overline{a}}]. \qquad \square$$

The next result gives a method of constructing elementary substructures of small cardinality. From this result it will follow that if a finite theory has a model, it has a countable model. In particular, if there is an infinite model of set theory, there is a countable model of set theory. This is an important result in set theory. In Chapter 5, we shall give a method to construct elementary extensions of arbitraily large cardinality.

Theorem 2.4.10 (Downward Löwenheim–Skolem theorem) *Let M be a structure of L and $X \subset M$. Suppose L has at most κ nonlogical symbols, κ an infinite cardinal number. Then there is an elementary substructure N of M such that $X \subset N$ and the cardinality of N is at most $\max(\kappa, |X|)$, where $|X|$ denotes the cardinality of X.*

Proof. Essentially, our N will be the smallest subset of M containing X satsfying the following conditions:

(i) Each $c_M \in N$, where c is a constant symbol of L.
(ii) The set N is closed under f_M for every function symbol f of L.
(iii) Whenever a sentence of the form $\exists v\varphi$ is valid in M, there is an element $a \in N$ such that $M \models \varphi_v[i_a]$.

By induction on k, we shall define

$$N_0 \subset N_1' \subset N_1 \subset \ldots \subset N_k \subset N_k' \subset N_{k+1} \subset \ldots \subset M$$

such that each N_k' is a substructure of N and for every formula of the form $\exists v\varphi[v, v_1, \ldots, v_n]$ and every $\overline{a} \in N_k'^n$, if $M \models \exists v\varphi[v, i_{\overline{a}}]$, there is $b \in N_{k+1}$ such that $M \models \varphi[i_b, i_{\overline{a}}]$. Further, each N_k is of cardinality $\leq \max(\kappa, |X|)$.

Let N_0 be the smallest subset of M containing X that contains all c_M's and that is closed under all f_M's. Note that $|N_0| \leq \max(\kappa, |X|)$ and that N_0 is a substructure of M.

Suppose N_k has been defined such that $|N_k| \leq \max(\kappa, |X|)$. We define N_k' and N_{k+1} now. Let N_k' be the smallest subset of M containing N_k that is closed under all f_M's. Note that $|N_k'| \leq \max(\kappa, |X|)$.

Fix a formula of the form $\varphi[v, v_1, \ldots, v_n]$. Let ψ be the formula $\exists v\varphi$. For $\overline{a} = (a_1, \ldots, a_n) \in (N_k')^n$, whenever

$$M \models \psi[i_{\overline{a}}],$$

there is $b \in M$ such that

$$M \models \varphi[i_b, i_{\overline{a}}].$$

Choose and fix one such b. Let N_{k+1} be obtained from N_k' by adding all the b's so chosen. Again note that $|N_{k+1}| \leq \max(\kappa, |X|)$.

Set

$$N = \cup_k N_k.$$

Then

(i) For every constant symbol c, $c_M \in N$.
(ii) For every function symbol f, N is closed under f_M.
(iii) $|N| \leq \max(\kappa, |X|)$ and $X \subset N$.

Thus, N is a substructure of M as in Remark 2.4.1.

Let $\varphi[v_1, \ldots, v_n]$ be any formula and $\overline{a} \in N^n$. Since N is a substructure of M, by Theorem 2.4.9, the proof will be complete if we show that for every formula $\varphi[v, v_1, \ldots, v_n]$ and for every $\overline{a} \in N^n$, if there is $b \in M$ satsfying

$$M \models \varphi[i_b, i_{\overline{a}}],$$

then there is $b \in N$ satsfying

$$M \models \varphi[i_b, i_{\overline{a}}].$$

Let $\varphi[v, v_1, \ldots, v_n]$ be a formula, and let $\overline{a} \in N^n$ and $b \in M$ be such that

$$M \models \varphi[i_b, i_{\overline{a}}].$$

Since $N_k \subset N_{k+1}$ for all k, there is a natural number p such that each $a_i \in N_p$. By the definition of N_{p+1} there is $b \in N_{p+1} \subset N$ such that

$$M \models \varphi[i_b, i_{\overline{a}}].$$

□

Remark 2.4.11 In the above proof we have used the following version of the axiom of choice.

Axiom of choice: If $\{X_i : i \in I\}$ is a family of nonempty sets, there is a map $f : I \to \cup_{i \in I} X_i$ such that $f(i) \in X_i$ for all $i \in I$.

A function f satsfying the conclusion of the axiom of choice is called a **choice function** for the family $\{X_i : i \in I\}$. The axiom of choice asserts only the existence of a choice function—it gives no method to produce a choice function.

From now on, by set theory we shall mean the theory obtained by adding the axiom of choice to the axioms of ZF. We denote this theory by ZFC.

Any model of the theory OF is called an ordered field. Let $\mathbb{F}$ be an ordered field and $A \subset \mathbb{F}$. An element u of $\mathbb{F}$ is called an **upper bound of** A if for every $a \in A$, $a \leq u$, where $x \leq y$ means that either $x < y$ or $x = y$. If u is an upper bound of A and no $v < u$ is an upper bound of A, then u is called the **least upper bound** of A. A **complete ordered field** is an ordered field $\mathbb{F}$ such that every nonempty subset A of $\mathbb{F}$ that has an upper bound has a least upper bound. It is known that if $\mathbb{F}_1$ and $\mathbb{F}_2$ are two complete ordered fields, then they are isomorphic, i.e., there is a field isomorphism $f : \mathbb{F}_1 \to \mathbb{F}_2$ such that for every $a, b \in \mathbb{F}_1$, $a < b \to f(a) < f(b)$. We denote a complete ordered field by $\mathbb{R}$. This is the traditional set of real numbers and is unique modulo the isomorphism just described. Cantor proved that $\mathbb{R}$ is uncountable.

Theorem 2.4.12 *There is no first-order theory with language L for ordered fields whose models are precisely complete ordered fields.*

Proof. This is because $\mathbb{R}$ is uncountable, and by the downward Löwenheim–Skolem theorem, any such theory must have a countable model. □

Remark 2.4.13 There is no set of first-order formulas of the language L for ordered fields expressing the least upper bound axiom: *Every nonempty set of real numbers that has an upper bound has a least upper bound.*

Let G be an abelian group. For any element $x \in G$, let nx denote the term

$$\underbrace{x + \cdots + x}_{n \text{ times}}.$$

We call a group G **divisible** if for every $n \geq 1$ and every $x \in G$ there exists a $y \in G$ such that $ny = x$. Call G **torsion-free** if for every $x \in G$, $x \neq 0$, and every $n \geq 1$, $nx \neq 0$.

Exercise 2.4.14 1. Show that there is a sequence of formulas of the language for group theory expressing that an abelian group is divisible. Later we shall show that such a finite set of formulas does not exist.
2. Show that there is a sequence of formulas of the language for group theory expressing that an abelian group is torsion-free. Later we shall show that such a finite set of formulas does not exist.

Remark 2.4.15 1. An ordered field $(A; 0, 1; +, \cdot; <)$ is called an **archimedean ordered field** if for every $x, y \in A$, $0 < x, y$, there is an n such that $y < nx$. Later we shall show that there is no first-order theory whose language is the same as the language for ordered fields, and whose models are precisely archimedean ordered fields.
2. A linearly ordered set $(A, <)$ is called **well-ordered** if for every nonempty subset B of A there is an element $b_0 \in B$ such that for every $b \in B$, $b_0 < b$, i.e., every nonempty subset of A has a least element. Using the axiom of choice, it is easy to show that a linearly ordered set is well-ordered if and only there is a sequence $\{a_n\}$ in A such that $a_{n+1} < a_n$ for all n. The statement "*Every nonempty set can be well-ordered*" is known as the **well-ordering principle**. It is known that the well-ordering principle is equivalent to the axiom of choice. Later we shall show that there is no first-order theory T whose language has only one nonlogical symbol—a binary relation symbol—and whose models are precisely well-ordered sets.

3

Propositional Logic

We now turn our attention to the fundamental notion of a logical deduction or a proof. As mentioned earlier, in mathematical parlance, statements valid in all models of a theory are called the theorems of the theory. But any statement is a well-formed finite sequence of symbols of the language. So, it is natural to expect a finitary definition of a theorem depending only on its subformulas and the syntactical construction. Note that while computing the truth value of a statement in a structure, one uses some rules of inference depending only on the syntactical construction of the statement. For instance, if A or B is true in a structure, we infer that $A \vee B$ is true in the structure. We have also noted that statements with some specific syntactical structures are valid in all structures. For instance, a statement of the form $\neg A \vee A$ is true in all structures. Statements true in all structures of a language are called *tautologies*. So all tautologies ought to be theorems. Is there a conveniently nice list of tautologies (to be called *logical axioms*) and a list of *rules of inference* such that a statement is valid if and only if it can be inferred from logical and nonlogical axioms using the rules of inference from our list? It is indeed the case.

In this chapter we first develop a simpler but important form of logic called *propositional logic*. The main objective of propositional logic is to formalize reasoning involving logical connectives $\vee$ and $\neg$ only.

A language for propositional logic has a nonempty set of variables and logical symbols $\vee$ and $\neg$ alone. Here variables stand for propositions rather than elements of a set. Then we consider the smallest set of expressions $\mathcal{F}$, to be called *formulas*, of the language that contains all variables and that is closed under $\neg$ and $\vee$. For instance, our set of variables could be the set of all elementary formulas of a first-order language L. Then $\mathcal{F}$ will coincide with the set of all formulas of L. We may think of variables of L to be some simple statements and formulas that can be made using variables and logical connectives. In this setting it is more important to examine the truth of a formula in terms of its subformulas.

In this chapter we shall introduce propositional logic in details.

3.1 Syntax of Propositional Logic

Thus *language for a propositional logic* L consists of

(i) **variables**: a nonempty set of symbols, and
(ii) **logical connectives**: $\neg$ and $\vee$.

Throughout this chapter, unless otherwise stated, L will denote the language of a propositional logic. A finite sequence of symbols of L will be called an expression in L.

Let $\mathcal{F}$ be the smallest set of expressions in L that contains all variables and that contains the expression $\neg A$ whenever $A \in \mathcal{F}$, and contains $\vee AB$ whenever A and B are in $\mathcal{F}$. The expressions belonging to $\mathcal{F}$ are called **formulas** of L.

Example 3.1.1 Let R be a binary relation on a nonempty set X. Let A stand for the proposition "R is reflexive," B for "R is symmetric," C for "R is transitive," and E for the proposition "R is an equivalence relation." Let the set of all variables of L be $\{A, B, C, E\}$. Then $E \leftrightarrow (A \wedge B \wedge C)$ is a formula of L standing for the statement "R is an equivalence relation if and only if R is reflexive, symmetric, and transitive."

Example 3.1.2 Let A stand for the statement "Humidity is high," B for "It will rain this afternoon," and C for "It will rain this evening." Let A, B, C be all the variables of L. Then the formula $A \rightarrow (B \vee C)$ stands for the statement "If humidity is high, it will rain this afternoon or this evening."

Exercise 3.1.3 Express the following statements as formulas of a propositional logic:

1. If the prime rateinterest goes up, people are not happy.
2. If stock prices go up, people are happy.

By a **literal** of L we shall mean either a variable or the negation of a variable of L. The rank of a formula is defined as in Chapter 1. As in the case of a first-order language, we shall often write $A \vee B$ for $\vee AB$. We shall maintain the same convention in using parentheses. Also, logical connectives $\wedge$, $\rightarrow$, and $\leftrightarrow$ are defined similarly. We shall use letters A, B, C, P, Q, R, and S, with or without subscripts for variables of L.

Let A be a formula of L. The set of all **subformulas** of A is the smallest set $\mathcal{S}$ of expressions in L satisfying the following conditions:

(i) $A \in \mathcal{S}$.
(ii) $A \in \mathcal{S}$ whenever $\neg A \in \mathcal{S}$.
(iii) $A, B \in \mathcal{S}$ whenever $A \vee B \in \mathcal{S}$.

The following example, mentioned in the beginning of this chapter, is an important example for us.

Example 3.1.4 Let L be a language for a first-order theory and L' the propositional logic whose variables are elementary formulas of L. Then the set of all formulas of L' is the same as the set of all formulas of L.

3.2 Semantics of Propositional Logic

In the next step we use the intuitive meaning of the logical connectives and define the truth or falsity of a formula in terms of its subformulas.

A **truth valuation** or an **interpretation** or a **structure** of L is a map v from the set of all variables of L to $\{T, F\}$.

Let v be an interpretation of L. We extend v (and denote the extension by v itself) to the set of all formulas by induction as follows:

$$v(\neg A) = T \text{ if and only if } v(A) = F$$

and

$$v(A \vee B) = T \text{ if and only if } v(A) = T \text{ or } v(B) = T.$$

If $v(A) = T$, we say that A is **true** in the structure v or that v satisfies A. Otherwise, A is said to be **false** in the structure.

Note that the truth value $v(A)$ of a formula A depends only on the variables occurring in A.

Exercise 3.2.1 Let A, B, C be all the variables of L. For each truth valuation v of L, compute the truth values of the following formulas:

1. $A \rightarrow B \rightarrow C$.
2. $B \rightarrow A \rightarrow C$.
3. $A \rightarrow C \rightarrow B$.
4. $(\neg A \vee B) \rightarrow \neg(A \wedge \neg B)$.

Let $\mathcal{A}$ be a set of formulas of L. An interpretation v is called a **model** of $\mathcal{A}$ if every $A \in \mathcal{A}$ is true in v. In this case we write $v \models \mathcal{A}$. If $\mathcal{A}$ has a model, we say that $\mathcal{A}$ is **satisfiable**.

Note that if a set of formulas is satisfiable, all its subsets are satisfiable.

Exercise 3.2.2 Let A, B be formulas and v a truth valuation of L. Show the following:

(a) $v(A \wedge B) = T$ if and only if $v(A) = v(B) = T$.
(b) $v(A \to B) = T$ if and only if $v(A) = F$ or $v(B) = T$.
(c) $v(A \leftrightarrow B) = T$ if and only if $v(A) = v(B)$.

Let A, B be formulas and $\mathcal{A}$ a set of formulas.

(i) We say that A is a **tautological consequence** of $\mathcal{A}$, and write $\mathcal{A} \models A$ if A is true in every model v of $\mathcal{A}$.
(ii) If A is a tautological consequence of the empty set of formulas, we say that A is a **tautology** and write $\models A$. Thus, A is a tautology if and only if $v(A) = T$ for every truth valuation v of L.
(iii) If $A \leftrightarrow B$ is a tautology (i.e., if $v(A) = v(B)$ for all truth valuations v), we say that A and B are **tautologically equivalent** and write $A \equiv B$.

Exercise 3.2.3(a) Show that $A \to B \to C$ and $B \to A \to C$ are tautologically equivalent.
(b) Show that $A \to B \to C$ and $A \to C \to B$ are not tautologically equivalent.
(c) Show that $A \to B \to C$ and $(A \to B) \to C$ are not tautologically equivalent.
(d) Show that $\neg(A \vee B) \vee C$ is a tautology if and only if both $\neg A \vee C$ and $\neg B \vee C$ are tautologies.
(e) Show that $\equiv$ is an equivalence relation on the set of all formulas of L.

The following result is quite easy to prove. Therefore, its proof is left as an exercise.

Proposition 3.2.4 *Let A, A_1, A_2, ..., A_n be formulas. Then the following statements are equivalent:*

(a) A is a tautological consequence of $A_1, A_2, \ldots, A_n$.
(b) $A_1 \to A_2 \to \cdots \to A_n \to A$ is a tautology.

Exercise 3.2.5(a) Show that

$$\models A \leftrightarrow \neg\neg A.$$

(b) Show that

$$\models \neg(A \vee B) \leftrightarrow \neg A \wedge \neg B.$$

(c) Show that

$$\models \neg(A \wedge B) \leftrightarrow \neg A \vee \neg B.$$

A formula is said to be in **conjunctive normal form** if it is a conjunction of disjunctions of literals, i.e., it is of the form

$$\wedge_{i=1}^{k} \vee_{j=1}^{n_k} A_{ij},$$

where each A_{ij} is a literal.

Exercise 3.2.6 Show that for every formula A there is a formula B in conjunctive normal form such that A and B are tautologically equivalent.

Proposition 3.2.7 *Let $A \to A^*$ be a map from the set of formulas of L to itself such that for all formulas A and B,*

$$(\neg A)^* = \neg(A^*) \text{ and } (A \vee B)^* = A^* \vee B^*.$$

Assume that B is a tautological consequence of $A_1, \ldots, A_n$. Then B^* is a tautological consequence of $A_1^*, \ldots, A_n^*$.

Proof. Let v be a truth valuation. For every variable A, define

$$v'(A) = v(A^*).$$

It is routine to check that for every formula A,

$$v'(A) = v(A^*).$$

Since B is a tautological consequence of $A_1, \ldots, A_n$, it is easy to see that $v(B^*) = T$ if $v(A_1^*) = \ldots = v(A_n^*) = T$. □

Exercise 3.2.8 Show that $\{A, \neg(A \vee B)\}$, $\{\neg A, \neg B, A \vee B\}$ are not satisfiable.

3.3 Compactness Theorem for Propositional Logic

We now present a nontrivial fact: *the compactness theorem for propositional logic.*

A binary relation $\leq$ on a set $\mathbb{P}$ is called a *partial order* if it is reflexive, antisymmetric, and transitive. An element p of $\mathbb{P}$ is called an *upper bound* of a subset A of $\mathbb{P}$ if $q \leq p$ for every $q \in A$. An element p of $\mathbb{P}$ is called a *maximal element* of $\mathbb{P}$ if for any $q \in \mathbb{P}$, $p \leq q$ implies that $p = q$. A partial order $\leq$ on $\mathbb{P}$ is called a linear order if for every $p, q \in \mathbb{P}$, $p \leq q$ or $q \leq p$. A *chain* in a partially ordered set $\mathbb{P}$ is a subset $\mathbb{Q}$ of $\mathbb{P}$ such that for every $p, q \in \mathbb{Q}$, $p \leq q$ or $q \leq p$, i.e., the restriction of $\leq$ to $\mathbb{Q}$ is a linear order.

Lemma 3.3.1 (Zorn's lemma) *Let $(\mathbb{P}, \leq)$ be a nonempty partially ordered set such that every chain in $\mathbb{P}$ has an upper bound in $\mathbb{P}$. Then $\mathbb{P}$ has a maximal element.*

Remark 3.3.2 It can be proved that the axiom of choice and Zorn's lemma are equivalent in ZF. In particular, Zorn's lemma is a theorem of ZFC.

We call $\mathcal{A}$ **finitely satisfiable** if every finite subset of $\mathcal{A}$ is satisfiable.

Clearly if $\mathcal{A}$ is satisfiable, it is finitely satisfiable. The compactness theorem tells us that the converse is also true. We proceed to prove this important result now.

Lemma 3.3.3 *Let $\mathcal{A}$ be a finitely satisfiable set of formulas and A a formula of L. Then either $\mathcal{A} \cup \{A\}$ or $\mathcal{A} \cup \{\neg A\}$ is finitely satisfiable.*

Proof. Suppose $\mathcal{A} \cup \{A\}$ is not finitely satisfiable. We need to show that for every finite subset $\mathcal{B}$ of $\mathcal{A}$, $\mathcal{B} \cup \{\neg A\}$ is satisfiable. Suppose this is not the case for a finite $\mathcal{B} \subset \mathcal{A}$. Then A is a tautological consequence of $\mathcal{B}$. Fix a finite $\mathcal{C} \subset \mathcal{A}$. Since $\mathcal{B} \cup \mathcal{C}$ is finite, it is satisfiable. Let v be a truth valuation that satisfies it. Since A is a tautological consequence of $\mathcal{B}$, $v(A) = T$. In particular, $\mathcal{C} \cup \{A\}$ is satisfiable. Thus, $\mathcal{A} \cup \{A\}$ is finitely satisfiable and we have arrived at a contradiction. □

Theorem 3.3.4 (Compactness theorem for propositional logic) *A set $\mathcal{A}$ of formulas of L is satisfiable if and only if it is finitely satisfiable.*

Proof. We need to prove the if part of the result only. Assume that $\mathcal{A}$ is finitely satisfiable. We have to show that $\mathcal{A}$ is satisfiable. Let $\mathbb{P}$ denote the set of all finitely satisfiable collections of formulas containing $\mathcal{A}$. Equip $\mathbb{P}$ with the partial order $\subset$ (contained in). Let $\mathbb{Q}$ be a chain in $\mathbb{P}$. Clearly, $\cup\mathbb{Q}$ is finitely satisfiable. So, it is an upper bound of $\mathbb{Q}$ in $\mathbb{P}$. Thus, by Zorn's lemma, there is a maximal family $\mathcal{M}$ of finitely satisfiable formulas containing $\mathcal{A}$. By Lemma 3.3.3, for every formula A, either $A \in \mathcal{M}$ or $\neg A \in \mathcal{M}$.

Consider the truth valuation v defined by

$$v(A) = T \Longleftrightarrow A \in \mathcal{M}, \qquad (*)$$

where A is a variable. We extend v to the set of all formulas such that for all formulas A and B,

$$v(\neg A) = T \Longleftrightarrow v(A) = F$$

and

$$v(A \vee B) = T \Longleftrightarrow v(A) \text{ or } v(B) = T.$$

By induction on the length of formulas A, we now show that $(*)$ holds for all formulas A. Since $\mathcal{A} \subset \mathcal{M}$, it will follow that $v(A) = T$ for every $A \in \mathcal{A}$ and the proof will be complete.

If A is a variable, the induction hypothesis holds for A by the definition of v.

Let A be of the form $\neg B$ and B satisfies $(*)$. Since $v(A) = T$ if and only if $v(B) = F$ and B satisfies $(*)$, $v(A) = T$ if and only if $B \notin \mathcal{M}$. But $B \notin \mathcal{M}$ if and only if $A \in \mathcal{M}$.

Finally, let A be a formula of the form $B \vee C$ and let B and C satisfy $(*)$. Let $v(A) = T$. Then either $v(B) = T$ or $v(C) = T$. So, either $B \in \mathcal{M}$ or $C \in \mathcal{M}$. If possible, suppose $B \vee C \notin \mathcal{M}$. Then $\neg(B \vee C) \in \mathcal{M}$. Since one of B and C belongs to $\mathcal{M}$, it follows that $\mathcal{M}$ is not finitely satisfiable.

Now assume that $v(A) = F$. Then $v(B) = v(C) = F$. So, $\neg B, \neg C \in \mathcal{M}$. If $A \in \mathcal{M}$, $\{B \vee C, \neg B, \neg C\}$ is satisfiable, which is impossible. Thus, $A \notin \mathcal{M}$. $\square$

Exercise 3.3.5 Assume that L is countable. Give a proof of the compactness theorem using induction on the natural numbers that does not require Zorn's lemma.

Hint: Since L is countable, the set of all its formulas is countable. Let $\{A_0, A_1, A_2, \ldots\}$ be an enumeration of all its formulas. We define a sequence of natural numbers $n_0, n_1, n_2, \ldots$ as follows: let n_0 be the first natural number i such that $A_i \notin \mathcal{A}$ and $\mathcal{A} \cup \{A_i\}$ is finitely satisfiable. Suppose n_i, $0 \leq i \leq k$, have been defined in such a way that $\mathcal{A} \cup \{A_{n_0}, \ldots, A_{n_k}\}$ is finitely satisfiable. Set $\mathcal{B} = \mathcal{A} \cup \{A_{n_i} : 0 \leq i \leq k\}$. Let n_{k+1} be the least natural number i such that $A_i \notin \mathcal{B}$ and $\mathcal{B} \cup \{A_i\}$ is finitely satisfiable, if such an i exists. Otherwise, set $n_{k+1} = n_k$. Now consider the set $\mathcal{M} = \mathcal{A} \cup \{A_{n_i} : i \in \mathbb{N}\}$.

Remark 3.3.6 One can prove the compactness theorem quite elegantly using Tychonoff's theorem for compact Hausdorff spaces. We give below the difficult part of the proof. Readers not familiar with these topics may skip the following proof.

Equip $X = \{T, F\}^{\mathcal{F}}$ with the product of discrete topologies on $\{T, F\}$, where $\mathcal{F}$ denotes the set of all variables of L. Thus, X is the set of all structures of L. By Tychonoff's theorem, X is compact Hausdorff.

For each finite $\mathcal{B} \subset \mathcal{A}$, set

$$F_{\mathcal{B}} = \{v \in X : v(A) = T \text{ for all } A \in \mathcal{B}\}.$$

It is easy to check that each $F_{\mathcal{B}}$ is closed in X. They are nonempty by hypothesis. By hypothesis again, the family

$$\{F_{\mathcal{B}} : \mathcal{B} \subset \mathcal{A} \text{ finite}\}$$

has the finite intersection property. Since X is compact, this implies that

$$\cap\{F_{\mathcal{B}} : \mathcal{B} \subset \mathcal{A} \text{ finite}\} \neq \emptyset.$$

Any v in this set models $\mathcal{A}$. □

As a corollary to the compactness theorem, we get the following useful result.

Proposition 3.3.7 *Let $\mathcal{A}$ be a set of formulas and A a formula. Then A is a tautological consequence of $\mathcal{A}$ if and only if A is a tautological consequence of a finite $\mathcal{B} \subset \mathcal{A}$.*

Proof. The if part of the result is clear. So, assume that for no finite $\mathcal{B} \subset \mathcal{A}$, $\mathcal{B} \models A$. It follows that $\mathcal{A} \cup \{\neg A\}$ is finitely satisfiable. Hence, it is satisfiable by Theorem 3.3.4. This implies that A is not a tautological consequence of $\mathcal{A}$. □

Exercise 3.3.8 A **graph** is an ordered pair $G = (V, E)$, where V is a nonempty set and E is a set of unordered pairs $\{x, y\}$ $(x \neq y)$ of elements of V. Elements of V are called **vertices** and those of E the **edges** of G. A **subgraph** of G is a graph $G' = (V', E')$, where $V' \subset V$ and $E' \subset E$. For any natural number $k \geq 1$, we say that G is k-**colorable** if there is a map $c : V \to \{1, 2, \ldots, k\}$ such that

$$\{x, y\} \in E \Rightarrow c(x) \neq c(y).$$

Show that a graph is k-colorable if and only if each of its finite subgraphs is k-colorable.

Hint: Let $G = (V, E)$ be a graph. Consider the language L for propositional logic with the set of variables

$$\{A_{xi} : x \in V, 1 \leq i \leq k\}.$$

Informally, we think of A_{xi} as the statement "the vertex x is assigned the color i." Now consider the set Φ of formulas consisting of the following formulas:

$$A_{x1} \vee \ldots \vee A_{xk}, \ \ x \in V,$$

$$\neg(A_{xi} \wedge A_{xj}), \ \ x \in V, 1 \leq i < j \leq k,$$

and

$$\neg(A_{xi} \wedge A_{yi}), \ \ \{x, y\} \in E, 1 \leq i \leq k.$$

Note that Φ is satisfiable means that G is k-colorable.

3.4 Proof in Propositional Logic

In this section we define a proof in propositional logic. To define a proof syntactically, we fix some tautologies and call them **logical axioms**. Further, we fix some **rules of inference**. There is only one class of logical axioms, called **propositional axioms**. These are formulas of the form $\neg A \vee A$.

Rules of inference of the propositional logic are

(a) **Expansion Rule.** Infer $B \vee A$ from A.
(b) **Contraction Rule.** Infer A from $A \vee A$.
(c) **Associative Rule.** Infer $(A \vee B) \vee C$ from $A \vee (B \vee C)$.
(d) **Cut Rule.** Infer $B \vee C$ from $A \vee B$ and $\neg A \vee C$.

Exercise 3.4.1 Show that the conclusion of any rule of inference is a tautological consequence of its hypotheses.

Let $\mathcal{A}$ be a set of formulas not containing any logical axiom. Elements of $\mathcal{A}$ will be called **nonlogical axioms**. A **proof** in $\mathcal{A}$ is a finite sequence of formulas $A_1, A_2, \ldots, A_n$ such that each A_i is either a logical axiom or a nonlogical axiom or can be inferred from formulas A_j, $j < i$, using one of the rules of inference. In this case we call the above sequence a proof of A_n in $\mathcal{A}$. If A has a proof in $\mathcal{A}$, we say that A is a **theorem** of $\mathcal{A}$ and write $\mathcal{A} \vdash A$. We call $\mathcal{A}$ **inconsistent** if there is a formula A such that $\mathcal{A} \vdash A$ and $\mathcal{A} \vdash \neg A$; $\mathcal{A}$ is called **consistent** if it is not inconsistent.

We shall write $\vdash A$ instead of $\emptyset \vdash A$. Note that each logical and nonlogical axiom is a theorem.

Remark 3.4.2 It is worth noting that there is an algorithm to decide whether a finite sequence of formulas is a proof, provided there is an algorithm to decide whether a formula is a nonlogical axiom.

Lemma 3.4.3 *If there is a sequence $A_1, A_2, \ldots, A_n$ such that each A_i is either a theorem of $\mathcal{A}$ or can be inferred from formulas A_j, $j < i$, using one of the rules of inference, then A_n is a theorem of $\mathcal{A}$.*

Proof. In the sequence $A_1, A_2, \ldots, A_n$, replace each A_i that is a theorem by a proof of it. The sequence thus obtained is a proof of A_n. □

Another easy but useful result is the following.

Lemma 3.4.4 *Let $\mathcal{A}$ be a set of formulas of L and A a formula of L. Suppose $\mathcal{A} \vdash A$. Then there is a finite $\mathcal{B} \subset \mathcal{A}$ such that $\mathcal{B} \vdash A$.*

Proof. This follows from the fact that each proof is a finite sequence of formulas and so contains only finitely many nonlogical axioms. □

Theorem 3.4.5 *Let L be the language of a propositional logic and A a formula of L. Then*

$$\vdash A \Longrightarrow \models A.$$

Proof. Let $A_1, A_2, \ldots, A_n$ be a proof of A. Thus, $A = A_n$. Fix any truth valuation v. By induction on i, $1 \leq i \leq n$, we show that $v(A_i) = T$. Note that A_1 must be a tautology. So, $v(A_1) = T$. Let $1 \leq i \leq n$ and $v(A_j) = T$ for all $j < i$. If A_i is a tautology, then $v(A_i) = T$. Otherwise, by Exercise 3.4.1, A_i is a tautological consequence of $\{A_j : j < i\}$. Hence $v(A_i) = T$. □

Exactly the same proof proves the following theorem.

Theorem 3.4.6 (Soundness theorem for propositional logic) *If $\mathcal{A}$ is a set of formulas of L and A a formula, then*

$$\mathcal{A} \vdash A \Longrightarrow \mathcal{A} \models A.$$

3.5 Metatheorems in Propositional Logic

Metatheorems in their own right are not very interesting. Their proofs are often mechanical and sometimes quite tedious. But they are unavoidable. Metatheorems play a key role in proving important results. In this section we prove some metatheorems in propositional logic. These results are needed to show that $\mathcal{A} \vdash A \Longleftrightarrow \mathcal{A} \models A$. This is a very important result because this shows that we have indeed formalized the notion of logical deduction in propositional logic.

Lemma 3.5.1 *Let A and B be formulas of L such that $\vdash A \vee B$. Then $\vdash B \vee A$.*

Proof. Consider the sequence

$$A \vee B, \quad \neg A \vee A, \quad B \vee A.$$

The first element of it is a theorem by the hypothesis, the second one is a propositional axiom, and the third one follows from the first two by the cut rule. By Lemma 3.4.3, $\vdash B \vee A$. □

Lemma 3.5.2 *A set of formulas $\mathcal{A}$ is inconsistent if and only if for every formula A, $\mathcal{A} \vdash A$.*

Proof. Let $\mathcal{A}$ be inconsistent. So there is a formula A such that both A and $\neg A$ are theorems of $\mathcal{A}$. Now take any formula B. By the expansion rule, $\vdash B \vee A$ and $\vdash B \vee \neg A$. By Lemma 3.5.1, $\vdash A \vee B$ and $\vdash \neg A \vee B$. By the cut rule, $\vdash B \vee B$. By the contraction rule, $\vdash B$. This proves the only if part of the result. The if part of the result is trivial. $\square$

Lemma 3.5.3 (Modus ponens) *Let A and B be formulas of L such that $\vdash A$ and $\vdash A \to B$. Then $\vdash B$.*

Proof. Since A is a theorem, by the expansion rule, $B \vee A$ is a theorem. Hence, by Lemma 3.5.1, $A \vee B$ is a theorem. Now consider the sequence

$$A \vee B, \quad A \to B, \quad B \vee B, \quad B.$$

The first formula is shown above to be a theorem; the second formula is a theorem by hypothesis; since $A \to B$ is the formula $\neg A \vee B$ by definition, the third formula is inferred from the first two by the cut rule; the last formula is inferred from the third formula by the contraction rule. The result follows by Lemma 3.4.3. $\square$

Now by induction on n, we easily get the following corollary.

Corollary 3.5.4 (Detachment rule) *Let $B, A_1, A_2, \ldots, A_n$ be formulas of L. Assume that each of $A_1, \ldots, A_n$ and $A_1 \to \cdots \to A_n \to B$ is a theorem. Then $\vdash B$.*

Lemma 3.5.5 *If $\vdash A \vee B$, then $\vdash \neg\neg A \vee B$.*

Proof. Since $\neg\neg A \vee \neg A$ is a logical axiom, $\vdash \neg\neg A \vee \neg A$. Then $\vdash \neg A \vee \neg\neg A$ by Lemma 3.5.1. By hypothesis, $\vdash A \vee B$. Hence $\vdash B \vee \neg\neg A$ by the cut rule. So, $\vdash \neg\neg A \vee B$ by Lemma 3.5.1. $\square$

Lemma 3.5.6 *Let $A_1, \ldots, A_n$ be formulas of L ($n \geq 2$) and $1 \leq i < j \leq n$. Suppose $\vdash A_i \vee A_j$. Then*

$$\vdash A_1 \vee \cdots \vee A_n.$$

Proof. We shall prove the result by induction on n. Clearly we can assume that $n \geq 3$.

If $i \geq 2$, then

$$\vdash A_2 \vee \cdots \vee A_n$$

by the induction hypothesis. Hence,

$$\vdash A_1 \vee A_2 \vee \cdots \vee A_n$$

by the expansion rule.

Now let $i = 1$ and $j \geq 3$. Then

$$\vdash A_1 \vee A_3 \vee \cdots \vee A_n$$

by the induction hypothesis. So,

$$\vdash (A_3 \vee \cdots \vee A_n) \vee A_1$$

by Lemma 3.5.1. So,

$$\vdash A_2 \vee ((A_3 \vee \cdots \vee A_n) \vee A_1)$$

by the expansion rule. Then

$$\vdash (A_2 \vee (A_3 \vee \cdots \vee A_n)) \vee A_1$$

by the associative rule. Hence,

$$\vdash A_1 \vee \cdots \vee A_n$$

by Lemma 3.5.1.

Finally, assume that $i = 1$ and $j = 2$. Then

$$\vdash (A_3 \vee \cdots \vee A_n) \vee (A_1 \vee A_2)$$

by the expansion rule. By the associative rule,

$$\vdash ((A_3 \vee \cdots \vee A_n) \vee A_1) \vee A_2.$$

By Lemma 3.5.1,

$$\vdash A_2 \vee ((A_3 \vee \cdots \vee A_n) \vee A_1).$$

By the associative rule,

$$\vdash (A_2 \vee \cdots \vee A_n) \vee A_1.$$

By Lemma 3.5.1 again,

$$\vdash A_1 \vee \cdots \vee A_n.$$

□

Lemma 3.5.7 *Let* $m \geq 1$, $n \geq 1$, *and* $1 \leq i_1, i_2, \ldots, i_m \leq n$. *Suppose*

$$\vdash A_{i_1} \vee A_{i_2} \vee \cdots \vee A_{i_m}.$$

Then

$$\vdash A_1 \vee A_2 \vee \cdots \vee A_n.$$

Proof. We shall prove the result by induction on m. In the rest of the proof, A will designate the formula $A_1 \vee \cdots \vee A_n$.

Case 1: $m = 1$. Set $i = i_1$. By hypothesis, $\vdash A_i$. By the expansion rule,

$$\vdash (A_{i+1} \vee \cdots \vee A_n) \vee A_i.$$

By Lemma 3.5.1,

$$\vdash A_i \vee A_{i+1} \vee \cdots \vee A_n.$$

Applying the expansion rule repeatedly, we get

$$\vdash A_1 \vee \cdots \vee A_n.$$

Case 2: $m = 2$. Suppose $i_1 = i_2$. Then $\vdash A_{i_1}$ by the hypothesis and the contraction rule. The result now follows from Case 1.

Suppose $i_2 < i_1$. Then, by the hypothesis and Lemma 3.5.1, $\vdash A_{i_2} \vee A_{i_1}$. Hence, without loss of generality, we assume that $i_1 < i_2$. The result now follows from Lemma 3.5.6.

Case 3: $m > 2$. By the hypothesis and the associative law,

$$\vdash (A_{i_1} \vee A_{i_2}) \vee A_{i_3} \vee \cdots \vee A_{i_m}.$$

Note that m is reduced by 1. Hence, by the induction hypothesis,

$$\vdash (A_{i_1} \vee A_{i_2}) \vee A.$$

By Lemma 3.5.1,

$$\vdash A \vee (A_{i_1} \vee A_{i_2}).$$

By the associative law,

$$\vdash (A \vee A_{i_1}) \vee A_{i_2}.$$

By the induction hypothesis,

$$\vdash (A \vee A_{i_1}) \vee A.$$

By Lemma 3.5.1,

$$\vdash A \vee (A \vee A_{i_1}).$$

By the associative law,

$$\vdash (A \vee A) \vee A_{i_1}.$$

By the induction hypothesis,

$$\vdash (A \vee A) \vee A.$$

By the induction hypothesis,

$$\vdash (A \vee A) \vee (A \vee A).$$

Applying the contraction rule twice, we see that

$$\vdash A.$$

□

Lemma 3.5.8 *If* $\vdash \neg A \vee C$ *and if* $\vdash \neg B \vee C$, *then* $\vdash \neg(A \vee B) \vee C$.

Proof. Since $\neg(A \vee B) \vee (A \vee B)$ is a propositional axiom, by Lemma 3.5.7,

$$\vdash A \vee B \vee \neg(A \vee B).$$

Since $\vdash \neg A \vee C$, by the cut rule,

$$\vdash (B \vee \neg(A \vee B)) \vee C.$$

By Lemma 3.5.1,

$$\vdash C \vee B \vee \neg(A \vee B).$$

Hence, by Lemma 3.5.7,

$$\vdash B \vee C \vee \neg(A \vee B).$$

Since $\vdash \neg B \vee C$, by the cut rule,

$$\vdash (C \vee \neg(A \vee B)) \vee C.$$

By Lemma 3.5.1,

$$\vdash C \vee C \vee \neg(A \vee B).$$

Hence

$$\vdash \neg(A \vee B) \vee C$$

by Lemma 3.5.7. □

3.6 Post Tautology Theorem

In this section we prove the following theorem due, to Emil Post.

Theorem 3.6.1 (Post tautology theorem) *If* A *is a formula of* L, *then*

$$\vdash A \Longleftrightarrow \models A.$$

By Theorem 3.4.5, we only need to prove that

Every tautology is a theorem.

Note that if A is a tautology, so is $A \vee A$. By the contraction rule, our result will be proved if we show that every tautology of the from $A \vee A$ is a theorem. We shall prove a bit more.

Proposition 3.6.2 *Let $n \geq 2$, and $A_1 \vee \cdots \vee A_n$ be a tautology. Then $\vdash A_1 \vee \cdots \vee A_n$.*

Proof. Suppose each A_i is a literal, i.e., each A_i is either a variable or the negation of a variable. Since $A_1 \vee \cdots \vee A_n$ is a tautology, there is a variable B such that both B and $\neg B$ occur in the sequence $A_1, \ldots, A_n$. But $\vdash \neg B \vee B$. Hence the result follows by Lemma 3.5.7.

So, we assume that for some $1 \leq i \leq n$, A_i is not a literal. By Lemma 3.5.7, without loss of generality, we assume that A_1 is not a literal. Thus A_1 is a formula in one of the three forms: $B \vee C$ or $\neg\neg B$ or $\neg(B \vee C)$.

We shall complete the proof by induction on the sum of lengths of the A_i's.

Case 1. A_1 is of the form $B \vee C$: Then $B \vee C \vee A_2 \vee \cdots \vee A_n$ is a tautology. Hence, it is a theorem by the induction hypothesis. The result in this case follows by the associative rule.

Case 2. A_1 is of the form $\neg\neg B$: From the hypothesis, it follows that $B \vee A_2 \vee \cdots \vee A_n$ is a tautology. Hence it is a theorem by the induction hypothesis. The result in this case now follows from Lemma 3.5.5.

Case 3. A_1 is of the form $\neg(B \vee C)$: Assuming the hypothesis, it is easy to check that $\neg B \vee A_2 \vee \cdots \vee A_n$ and $\neg C \vee A_2 \vee \cdots \vee A_n$ are tautologies. So, by the induction hypothesis, they are theorems. The result in this case follows from Lemma 3.5.8. □

There is another very useful formulation of the Post tautology theorem.

Theorem 3.6.3 *If $\vdash A_1, \ldots, \vdash A_n$ and if B is a tautological consequence of $A_1, \ldots, A_n$, then $\vdash B$.*

Proof. By Lemma 3.2.4, $A_1 \to \cdots \to A_n \to B$ is a tautology. Hence, by the Post tautology theorem,

$$\vdash A_1 \to \cdots \to A_n \to B.$$

The result now follows from Corollary 3.5.4 (the detachment rule). □

Theorem 3.6.4 (Completeness theorem for propositional logic) *Let $\mathcal{A}$ be a set of formulas of L. Then*

$$\mathcal{A} \vdash A \Longleftrightarrow \mathcal{A} \models A.$$

Proof. Assume that $\mathcal{A} \vdash A$. Then there is a finite set of formulas $\{B_1, \ldots, B_n\}$ such that

$$\{B_1, \ldots, B_n\} \vdash A.$$

By the soundness theorem,

$$\{B_1, \ldots, B_n\} \models A.$$

Hence,

$$\mathcal{A} \models A.$$

Conversely, assume that $\mathcal{A} \models A$. Then, by the compactness theorem (Proposition 3.3.7), there is a finite set $\mathcal{B} \subset \mathcal{A}$ such that $\mathcal{B} \models A$. Hence, by Theorem 3.6.3, $\mathcal{B} \vdash A$. In particular, $\mathcal{A} \vdash A$.

Corollary 3.6.5(*a*) *A set of formulas $\mathcal{A}$ is consistent if and only if it is satisfiable.*

(*b*) *Suppose A and B are tautologically equivalent. Then $\vdash A$ if and only if $\vdash B$.*

(*c*) *Suppose $\vdash A \leftrightarrow B$. Then $\vdash A$ if and only if $\vdash B$.*

(*d*) *$\vdash A \rightarrow B$ if and only if $\vdash \neg B \rightarrow \neg A$.*

(*e*) *$\vdash A \wedge B$ if and only if $\vdash A$ and $\vdash B$.*

(*f*) *If $\vdash A \rightarrow B$ and $\vdash B \rightarrow C$, then $\vdash A \rightarrow C$.*

The simple proof of this result is left as an exercise.

Remark 3.6.6 The completeness theorem gives a trivial proof of the compactness theorem. To see this, let $\mathcal{A}$ be a set of formulas of L, and A a formula of L. Then

$$\begin{aligned} \mathcal{A} \models A &\Rightarrow \mathcal{A} \vdash A \\ &\Rightarrow \mathcal{B} \vdash A \quad \text{for some finite } \mathcal{B} \subset \mathcal{A} \\ &\Rightarrow \mathcal{B} \models A \quad \text{for some finite } \mathcal{B} \subset \mathcal{A}. \end{aligned}$$

The first implication holds by Theorem 3.6.4, the second implication holds because every proof contains only finitely many nonlogical axioms, and the last one holds by the soundness theorem for propositional logic.

4

Proof and Metatheorems in First-Order Logic

In the previous chapter, we introduced the notion of proof in propositional logic. In this chapter we shall define proof in a first-order theory and prove several metatheorems in first-order logic.

4.1 Proof in First-Order Logic

In Chapter 1, we have already described what a first-order language is and what its terms and formulas are. We fix a first-order language L.

The **logical axioms** of L are:

(a) **Propositional axioms**: These are formulas of the form $\neg A \vee A$.

(b) **Identity axioms**: These are formulas of the form $x = x$, where x is a variable;

(c) **Equality axioms**: These are formulas of the form

$$y_1 = z_1 \to \cdots \to y_n = z_n \to fy_1 \cdots y_n = fz_1 \cdots z_n$$

or formulas of the form

$$y_1 = z_1 \to \cdots \to y_n = z_n \to py_1 \cdots y_n \to pz_1 \cdots z_n;$$

(d) **Substitution axioms**: These are formulas of the form $A_x[t] \to \exists x A$, where A is a formula and t a term substitutable for x in A.

The **rules of inference** of L are:

(a) **Expansion rule.** Infer $B \vee A$ from A.
(b) **Contraction rule.** Infer A from $A \vee A$.
(c) **Associative rule.** Infer $(A \vee B) \vee C$ from $A \vee (B \vee C)$.
(d) **Cut rule.** Infer $B \vee C$ from $A \vee B$ and $\neg A \vee C$.
(e) **∃-Introduction rule**: If x is not free in B, infer $\exists x A \rightarrow B$ from $A \rightarrow B$.

Each logical axiom is valid in every structure of L. Also, the conclusion of each rule of inference is valid in any structure of L in which its hypotheses are valid.

Let T be a first-order theory. A **proof** in T, a theorem in T, etc., are defined as before. So, a **proof in** T is a finite sequence $A_1, \ldots, A_n$ of formulas of $L(T)$ such that for each $i \leq n$, A_i is either an axiom (logical or nonlogical) of $L(T)$ or can be inferred from $\{A_j : j < i\}$ by a rule of inference. We shall write $T \vdash A$ or simply $\vdash A$ (when T is understood) to say that A is a theorem of T.

We prove the soundness theorem (also known as the validity theorem) for first-order theories in exactly the same way we proved the soundness theorem for propositional logic (Theorem 3.4.6). Recall that a formula in a theory is valid if it is true in all models of the theory.

Theorem 4.1.1 (Validity theorem) *Every theorem in T is valid in T.*

A theory T' is called an **extension** of T if $L(T')$ is an extension of $L(T)$ and if every nonlogical axiom of T is a theorem of T'. If $L(T')$ is an extension of $L(T)$ and if every nonlogical axiom of T is a nonlogical axiom of T', then T is called a **part** of T'; moreover, if the number of nonlogical axioms of T is finite, it is called a **finitely axiomatized** part of T'. If T and T' are extensions of each other, then they are called **equivalent**. Note that if T and T' are equivalent, then they have the same language, i.e., $L(T) = L(T')$. Let Γ be a set of formulas of T. The simple extension of the theory T obtained by adding Γ as new nonlogical axioms is designated by $T[\Gamma]$.

Exercise 4.1.2 Let T' be an extension of T. Show that every theorem of T is a theorem of T'.

An extension T' of T is called a **conservative extension** of T if every formula of T that is a theorem of T' is also a theorem of T; T' is a **simple extension** of T if $L(T) = L(T')$ and every theorem of T is a theorem of T' too.

4.2 Metatheorems in First-Order Logic

The converse of the validity theorem is also true. This is a famous result that was first proved by Gödel and is known as the completeness theorem.

A proof of it will be given in the next chapter. In this section we present a few metatheorems needed to prove the completeness theorem.

Throughout this section, we fix a first-order theory T; by a theorem we shall mean a theorem of T, and $\vdash A$ will mean that A is a theorem of T.

The next few definitions are needed to adapt results from propositional logic to first-order logic. Recall that a formula is called elementary if it is either an atomic formula or a formula of the form $\exists xB$. We have already seen that the formulas of L are precisely the formulas of the language of the propositional logic whose variables are precisely the elementary formulas of L. A **truth valuation** for L is a map v from the set of all elementary formulas into $\{T, F\}$. We extend v to the set of all formulas of L as before. Further, we define the notion of **tautological consequences**, **tautology**, and **Tautologically equivalent formulas**, in exactly the same way as before. For instance, A is a tautological consequence of a set $\mathcal{A}$ of formulas of L if $v(A) = T$ for every truth valuation v of L in which all formulas of $\mathcal{A}$ are true.

It is also clear that the detachment rule (Theorem 3.6.3), the Post tautology theorem (Theorem 3.6.1) etc. proved in the last chapter hold for first-order theories also.

Proposition 4.2.1 (Detachment rule) *Suppose*

$$\vdash A_1, \ldots, \vdash A_n$$

and

$$\vdash A_1 \to \cdots \to A_{n-1} \to A_n \to A.$$

Then

$$\vdash A.$$

Theorem 4.2.2 (Post tautology theorem) *Suppose*

$$T \vdash A_1, \ldots, T \vdash A_n$$

and

$$A_1, \ldots, A_n \models A.$$

Then

$$T \vdash A.$$

Theorem 4.2.3 *Every tautology in a first-order theory is a theorem of the theory.*

The corollary to the tautology theorem given in the last chapter also holds for first-order theories. In the rest of this section, we shall prove some metatheorems involving terms, quantifiers, etc.

Lemma 4.2.4

$$\vdash A \to \exists v_1 \cdots \exists v_n A.$$

Proof. Applying the substitution axiom repeatedly, we get

$$\vdash A \rightarrow \exists v_n A,$$

$$\vdash \exists v_n A \rightarrow \exists v_{n-1} \exists v_n A,$$

$$\vdots$$

$$\vdash \exists v_2 \ldots \exists v_n A \rightarrow \exists v_1 \cdots \exists v_n A.$$

Since $A \rightarrow \exists v_1 \ldots \exists v_n A$ is a tautological consequence of the above formulas, the result follows from the tautology theorem. □

Proposition 4.2.5 (**∀-introduction rule**) *If* $\vdash A \rightarrow B$ *and* x *is not free in* A, *then* $\vdash A \rightarrow \forall x B$.

Proof. By the hypothesis and the tautology theorem, we have

$$\vdash \neg B \rightarrow \neg A.$$

Then

$$\vdash \exists x \neg B \rightarrow \neg A$$

by the ∃-introduction rule. So,

$$\vdash A \rightarrow \neg \exists x \neg B.$$

Hence,

$$\vdash A \rightarrow \forall x B$$

by the definition of ∀. □

Proposition 4.2.6 (**Generalization rule**) *If* $\vdash A$, *then* $\vdash \forall x A$.

Proof. By the hypothesis and the expansion rule,

$$\vdash \neg \forall x A \rightarrow A.$$

Then, by the ∀-introduction rule (Proposition 4.2.5),

$$\vdash \neg \forall x A \rightarrow \forall x A.$$

The result follows by the tautology theorem. □

Proposition 4.2.7 (**Substitution rule**) *If* B *is an instance of* A *and if* $\vdash A$, *then* $\vdash B$.

Proof. We first prove the result in a simple case. Suppose t is substitutable for v in A and B is $A_v[t]$. By the substitution axiom,

$$\vdash \neg A_v[t] \to \exists v \neg A.$$

So, by the tautology theorem,

$$\vdash \forall v A \to B.$$

By hypothesis and the generalization rule,

$$\vdash \forall v A.$$

Hence

$$\vdash B$$

by the detachment rule.

Now let B be the formula $A_{v_1,\dots,v_n}[t_1,\dots,t_n]$. Let $w_1,\dots,w_n$ be variables, each different from $v_1,\dots,v_n$, not occurring in A or B. By repeated application of the first part,

$$\vdash A_{v_1}[w_1],$$

$$\vdash A_{v_1,v_2}[w_1,w_2],$$

$$\vdots$$

$$\vdash A_{v_1,\dots,v_n}[w_1,\dots,w_n].$$

Let C designate the formula $A_{v_1,\dots,v_n}[w_1,\dots,w_n]$. Then B is the formula $C_{w_1,\dots,w_n}[t_1,\dots,t_n]$. By repeated application of the first part, we see that

$$\vdash C_{w_1}[t_1],$$

$$\vdash C_{w_1,w_2}[t_1,t_2],$$

$$\vdots$$

$$\vdash C_{w_1,\dots,w_n}[t_1,\dots,t_n].$$

□

Using Lemma 4.2.4, Proposition 4.2.6 and the substitution rule (Proposition 4.2.7), we get the following result.

Proposition 4.2.8 (Substitution theorem)

(a) $\vdash A_{v_1,\dots,v_n}[t_1,\dots,t_n] \to \exists v_1 \cdots \exists v_n A.$
(b) $\vdash \forall v_1 \cdots \forall v_n A \to A_{v_1,\dots,v_n}[t_1,\dots,t_n].$

Proposition 4.2.9 (Closure theorem) *Let B be the closure of A. Then $\vdash A$ if and only if $\vdash B$.*

Proof. If A is a theorem, then B is a theorem by the generalization rule (Proposition 4.2.6).

Since $A = A_{v_1,\ldots,v_n}[v_1, \ldots, v_n]$, by the substitution theorem, $\vdash B \to A$. So, if $\vdash B$, $\vdash A$ by the detachment rule. □

Exercise 4.2.10 Let $L(T')$ be an extension of $L(T)$. Show that the following statements are equivalent.

(i) The theory T' is a conservative extension of T.
(ii) A sentence of T is a theorem of T' if and only if it is a theorem of T.

Proposition 4.2.11 (Distribution rule) *If $\vdash A \to B$, then*

$$\vdash \exists v A \to \exists v B$$

and

$$\vdash \forall v A \to \forall v B.$$

Proof. By the substitution axiom,

$$\vdash B \to \exists v B.$$

By the hypothesis and the tautology theorem,

$$\vdash A \to \exists v B.$$

So,

$$\vdash \exists v A \to \exists v B$$

by the $\exists$-introduction rule.

By the substitution theorem,

$$\vdash \forall v A \to A.$$

By the hypothesis and the tautology theorem,

$$\vdash \forall v A \to B.$$

So,

$$\vdash \forall v A \to \forall v B$$

by the $\forall$-introduction rule. □

Formulas A and B are called **equivalent** in T if

$$T \vdash A \leftrightarrow B.$$

We write $A \equiv_T B$ if A and B are equivalent in T.

Exercise 4.2.12 Show that $\equiv_T$ is an equivalence relation on the set of all formulas of T.

Proposition 4.2.13 (Equivalence theorem) *Let A' be obtained from A by simultaneously replacing subformulas $B_1, \ldots, B_n$ by $B'_1, \ldots, B'_n$ respectively. Assume that*

$$\vdash B_i \leftrightarrow B'_i,$$

$1 \leq i \leq n$. *Then*

$$\vdash A \leftrightarrow A'.$$

Proof. If A is one of the formulas B_i, then there is nothing to prove. We assume that this is not the case and prove the result by induction on the length of A.

Let A be a formula of the form $\neg B$. Then A' is $\neg B'$, where B' is obtained by replacing subformulas $B_1, \ldots, B_n$ by $B'_1, \ldots, B'_n$ respectively. Then

$$\vdash B \leftrightarrow B'$$

by the induction hypothesis. So,

$$\vdash A \leftrightarrow A'$$

by the tautology theorem.

The proof is similar if A is a formula of the form $B \vee C$.

Now assume that A is a formula of the form $\exists v B$. Then by the induction hypothesis,

$$\vdash B \leftrightarrow B',$$

where B' is obtained by replacing subformulas $B_1, \ldots, B_n$ by $B'_1, \ldots, B'_n$ respectively. The result now follows from the distribution rule. □

Sometimes there is difficulty in substituting a term t for a variable v in a formula A. Our next result shows a way to circumvent this.

Let B be obtained from A by a sequence of replacements of the following type: replace a subformula $\exists v C$ by a formula of the form $\exists w C_v[w]$, where w is not free in C. Then B is called a **variant** of the formula A.

Proposition 4.2.14 (Variant theorem) *If B is a variant of A, then*

$$\vdash A \leftrightarrow B.$$

Proof. By the equivalence theorem and the tautology theorem, we only need to show that

$$\vdash \exists v C \leftrightarrow \exists w C_v[w], \tag{1}$$

where w is not free in C.

By the substitution axiom,

$$\vdash C_v[w] \to \exists v C.$$

So, by the $\exists$-introduction rule,

$$\vdash \exists w C_v[w] \to \exists v C. \tag{2}$$

On the other hand, since w is not free in C, the formula $(C_v[w])_w[v]$ is C. In other words, if we replace free occurrences of v in C by w and then replace free occurrences of w by v, we get back C because w is not free in C. Hence, by the substitution axiom,

$$\vdash C \to \exists w C_v[w]. \tag{3}$$

By the $\exists$-introduction rule,

$$\vdash \exists v C \to \exists w C_v[w]. \tag{4}$$

Since the formula in (1) is a tautological consequence of those in (2) and (4), (1) follows from the tautology theorem. □

Proposition 4.2.15 (Symmetry theorem) *For any two terms t and s,*

$$\vdash t = s \leftrightarrow s = t.$$

Proof. Let v and w be distinct variables. By the equality axioms,

$$\vdash v = w \to v = v \to v = v \to w = v.$$

By the identity axiom and the tautology theorem,

$$\vdash v = w \to w = v.$$

Substituting t for v and s for w, by the substitution rule,

$$\vdash t = s \to s = t$$

and

$$\vdash s = t \to t = s.$$

The result now follows from the tautology theorem. □

Proposition 4.2.16 (Equality theorem)

(a) Let a term s be obtained from t by replacing subterms $t_1, \ldots, t_n$ by $s_1, \ldots, s_n$ respectively. If

$$\vdash t_i = s_i,$$

$1 \le i \le n$, then

$$\vdash t = s.$$

(b) *Let a formula B be obtained from A by replacing some occurrences of terms $t_1, \ldots, t_n$ in A not immediately following $\exists$ or $\forall$ by $s_1, \ldots, s_n$ respectively. If*

$$\vdash t_i = s_i,$$

$1 \le i \le n$, *then*

$$\vdash A \leftrightarrow B.$$

Proof. (a) We shall prove the result by induction on the rank of t.

If t is t_i for some i, s is the term s_i and there is nothing to be proved. This, in particular, shows that the result is true for t of rank 0.

Let t be a term of the form $fa_1 \cdots a_k$. Then, the t_i's are subterms of the a_j's. (Why?) Let a'_j be the term obtained from a_j by replacing appropriate occurrences of terms $t_1, \ldots, t_n$ by $s_1, \ldots, s_n$ respectively. Then by the induction hypothesis,

$$\vdash a_j = a'_j,$$

$1 \le j \le k$. We have the equality axiom

$$x_1 = y_1 \to \cdots \to x_k = y_k \to fx_1 \cdots x_k = fy_1 \cdots y_k.$$

So, the result follows from the substitution rule and the tautology theorem.

(b) We prove the result by induction on the rank of A.

Assume A is an atomic formula of the form $pa_1 \cdots a_k$. Then the t_i's are subterms of the a_j's. Let a'_j be the term obtained from a_j by replacing appropriate occurrences of terms $t_1, \ldots, t_n$ by $s_1, \ldots, s_n$ respectively. Then B is the formula $pa'_1 \cdots a'_k$. By (a),

$$\vdash a_j = a'_j,$$

$1 \le j \le k$. We have the equality axiom

$$\vdash x_1 = y_1 \to \cdots \to x_k = y_k \to px_1 \cdots x_k \to py_1 \cdots y_k.$$

By the substitution rule and the tautology theorem,

$$\vdash pa_1 \cdots a_k \to pa'_1 \cdots a'_k.$$

Since by the symmetry theorem,

$$\vdash a'_j = a_j,$$

by the tautology theorem,

$$\vdash pa'_1 \cdots a'_k \to pa_1 \cdots a_k.$$

Hence,

$$\vdash pa_1 \cdots a_k \leftrightarrow pa'_1 \cdots a'_k$$

by the tautology theorem.

Let A be a formula of the form $\exists vC$. Then, by the hypothesis in (b), B is the formula $\exists vD$, where D is obtained from C by replacing appropriate occurrences of terms $t_1, \ldots, t_n$ by $s_1, \ldots, s_n$ respectively. By the induction hypothesis,

$$\vdash C \leftrightarrow D.$$

The result follows from the distribution rule and the tautology theorem. The cases in which A is of the form $\neg C$ or $C \vee D$ are dealt with similarly. □

In mathematics, while proving a sentence of the form $A \rightarrow B$, quite often one assumes A and then proves B. This means that one adds A as a new axiom and proves B in this extension of T. We now show that this is a correct method of proving $A \rightarrow B$ in T.

Recall that we designated the simple extension of a theory T obtained by adding a set Γ of formulas as new axioms by $T[\Gamma]$. If $\Gamma = \{A_1, \ldots, A_n\}$, we shall write $T[A_1, \ldots, A_n]$ instead of $T(\Gamma)$.

Proposition 4.2.17 (Deduction theorem) *Let A be a closed formula. Then*

$$T \vdash A \rightarrow B$$

if and only if

$$T[A] \vdash B.$$

Proof. Suppose

$$T \vdash A \rightarrow B.$$

Then

$$T[A] \vdash A \rightarrow B.$$

Also

$$T[A] \vdash A.$$

So,

$$T[A] \vdash B$$

by the detachment rule.

Now assume that

$$T[A] \vdash B.$$

Fix a proof of $A_1, \ldots, A_n$ of B in $T[A]$. By induction on i, we shall prove that

$$T \vdash A \rightarrow A_i$$

for $1 \leq i \leq n$, which will complete the proof.

The formula A_1 is an axiom of $T[A]$. If A_1 is an axiom of T, then

$$T \vdash A_1.$$

Hence,

$$T \vdash A \rightarrow A_1$$

by the expansion rule. In case A_1 is A,

$$T \vdash A \rightarrow A$$

by the propositional axiom.

Assume the hypothesis for all $j < i$. If A_i is an axiom of $T[A]$, we have already proved that

$$T \vdash A \rightarrow A_i.$$

If A_i is inferred from $\{A_j : j < i\}$ using a rule of inference other than the $\exists$-introduction rule, then A_i is a tautological consequence of $\{A_j : j < i\}$. But then $A \rightarrow A_i$ is a tautological consequence of $\{A \rightarrow A_j : j < i\}$. By the induction hypothesis,

$$T \vdash A \rightarrow A_j$$

for every $j < i$. Hence,

$$T \vdash A \rightarrow A_i$$

by the tautology theorem.

Now assume that A_i is inferred from some A_j, $j < i$, by the $\exists$-introduction rule. So, A_j is a formula of the form $B \rightarrow C$, A_i of the form $\exists v B \rightarrow C$, where v is not free in C. By the induction hypothesis,

$$T \vdash A \rightarrow B \rightarrow C;$$

by the tautology theorem,

$$T \vdash B \rightarrow A \rightarrow C.$$

Since A is closed, v is not free in $A \rightarrow C$. Hence by the $\exists$-introduction rule,

$$T \vdash \exists v B \rightarrow A \rightarrow C;$$

by the tautology theorem,

$$T \vdash A \rightarrow \exists v B \rightarrow C,$$

i.e.,

$$T \vdash A \rightarrow A_i.$$

□

Corollary 4.2.18 *If $A_1, \ldots, A_n$ are closed, then*

$$T[A_1, \ldots, A_n] \vdash B \Longleftrightarrow T \vdash A_1 \rightarrow \cdots \rightarrow A_n \rightarrow B.$$

Exercise 4.2.19 For any consistent theory T that has a model with more than one point, show that

$$T[x = y] \vdash \forall x \forall y (x = y)$$

but

$$T \nvdash x = y \to \forall x \forall y (x = y).$$

Thus, the deduction theorem is not true if the A's are not closed.

Proposition 4.2.20 (Theorem on constants) *Let T' be obtained from T by adding new constants but no new nonlogical axioms. Then T' is a conservative extension of T.*

Proof. By the definition of conservative extension, we need to show that for any formula φ of T,

$$T' \vdash \varphi \Rightarrow T \vdash \varphi.$$

Suppose

$$T' \vdash \varphi.$$

Hence, by the substitution rule,

$$T' \vdash \varphi_{v_1,\ldots,v_n}[c_1, \ldots, c_n],$$

where $v_1, \ldots, v_n$ are all the variables free in φ and $c_1, \ldots, c_n$ are new constants.

Fix a proof of $\varphi_{v_1,\ldots,v_n}[c_1, \ldots, c_n]$ in T' and replace all occurrences of $c_1, \ldots, c_n$ in the proof by distinct variables $w_1, \ldots, w_n$ respectively that do not occur in the proof. Note that we have thus obtained a proof of $\psi = \varphi_{v_1,\ldots,v_n}[w_1, \ldots, w_n]$ in T. Since $\varphi = \psi_{v_1,\ldots,v_n}[w_1, \ldots, w_n]$, the result now follows from the substitution rule. □

Proposition 4.2.21 *Let t, $t_1, \ldots, t_n$ and $s_1, \ldots, s_n$ be terms. Then*

$$\vdash t_1 = s_1 \to \cdots \to t_n = s_n \to t[t_1, \ldots, t_n] = t[s_1, \ldots, s_n].$$

Proof. Replace each variable occurring in a t_i or in an s_i by a new constant. Designate the extension of T thus obtained by T'. Let t_i become t'_i, s_i become s'_i, and t become t'. By the theorem on constants, it suffices to prove that

$$T' \vdash t'_1 = s'_1 \to \cdots \to t'_n = s'_n \to t'[t'_1, \ldots, t'_n] = t'[s'_1, \ldots, s'_n].$$

By the deduction theorem, this will follow from

$$T'[t'_1 = s'_1, \ldots, t'_n = s'_n] \vdash t'[t'_1, \ldots, t'_n] = t'[s'_1, \ldots, s'_n].$$

This follows from the equality theorem. □

In the same way we prove the following result.

Proposition 4.2.22 *Let $t_1, \ldots, t_n$ and $s_1, \ldots, s_n$ be terms and A a formula of L. Then*

$$\vdash t_1 = s_1 \rightarrow \cdots \rightarrow t_n = s_n \rightarrow (A[t_1, \ldots, t_n] \leftrightarrow A[s_1, \ldots, s_n]).$$

Proposition 4.2.23 *Let a variable v not occur in the term t and let A be a formula of L such that t is substitutable for v in A. Then*

$$\vdash A_v[t] \leftrightarrow \exists v(v = t \wedge A).$$

Proof. By Proposition 4.2.22,

$$\vdash v = t \rightarrow (A \leftrightarrow A_v[t]).$$

So, by the tautology theorem,

$$\vdash (v = t \wedge A) \rightarrow A_v[t].$$

By the $\exists$-introduction rule,

$$\vdash \exists v(v = t \wedge A) \rightarrow A_v[t].$$

On the other hand, by the substitution axiom,

$$\vdash (t = t \wedge A_v[t]) \rightarrow \exists v(v = t \wedge A).$$

Since $\vdash t = t$, by the tautology theorem, it follows that

$$\vdash A_v[t] \rightarrow \exists v(v = t \wedge A).$$

The result now follows from the tautology theorem. □

Exercise 4.2.24 Let T be a first-order theory. Let $\mathcal{O}$ be the smallest set of formulas that contains all atomic formulas and their negations and that is closed under disjunctions and conjunctions. Show that every open formula is equivalent in T to a formula in $\mathcal{O}$.

Exercise 4.2.25 Show that a formula of T of the form $\neg(fx_1 \cdots x_n = y)$ is equivalent in T to a formula of the form

$$\exists z(\neg(y = z) \wedge fx_1 \cdots x_n = z),$$

$y, z, x_1, \ldots, x_n$ distinct.

Exercise 4.2.26 (a) Show that

$$\vdash \neg\exists v A \leftrightarrow \forall v \neg A.$$

(b) Show that

$$\vdash \neg\forall v A \leftrightarrow \exists v \neg A.$$

(c) If v is not free in B, show that

$$\vdash \exists v A \vee B \leftrightarrow \exists v(A \vee B),$$

$$\vdash \forall v A \vee B \leftrightarrow \forall v(A \vee B),$$

$$\vdash B \vee \exists v A \leftrightarrow \exists v(B \vee A),$$

and

$$\vdash B \vee \forall v A \leftrightarrow \forall v(B \vee A).$$

Exercise 4.2.27 Show the following:

(a) $\vdash \exists v(A \vee B) \leftrightarrow \exists v A \vee \exists v B$.
(b) $\vdash \forall v(A \wedge B) \leftrightarrow \forall v A \wedge \forall v B$.
(c) $\vdash \exists v(A \wedge B) \rightarrow \exists v A \wedge \exists v B$.
(d) $\vdash \forall v A \vee \forall v B \rightarrow \forall v(A \vee B)$.

Exercise 4.2.28 Give examples A and B of formulas of N such that the formulas

$$\forall v(A \vee B) \rightarrow \forall v A \vee \forall v B$$

and

$$\exists v A \wedge \exists v B \rightarrow \exists v(A \wedge B)$$

are not theorems of N.

Exercise 4.2.29 Let v and w be distinct variables. Show the following:

(a) $\vdash \exists v \exists w A \leftrightarrow \exists w \exists v A$.
(b) $\vdash \forall v \forall w A \leftrightarrow \forall w \forall v A$.
(c) $\vdash \exists v \forall w A \rightarrow \forall w \exists v A$.

Exercise 4.2.30 Give a formula of N of the form

$$\forall v \exists w A \rightarrow \exists w \forall v A$$

that is not a theorem.

A formula A is said to be in **prenex form** if it is in the form

$$Q_1 v_1 \cdots Q_n v_n B,$$

where each Q_i is either $\exists$ or $\forall$, and B open. Then $Q_1 v_1 \cdots Q_n v_n$ is called the **prefix** and B the **matrix** of A. A formula in prenex form is called **existential** if all the quantifiers in its prefix are $\exists$; a formula in prenex form is called **universal** if all the quantifiers in its prefix are $\forall$.

Exercise 4.2.31 Show that every formula is equivalent in T to a formula in prenex form whose matrix is of the form

$$\wedge_{i=1}^{k} \vee_{j=1}^{n_i} B_{ij},$$

where each B_{ij} is either an elementary formula or the negation of an elementary formula.

4.3 Some Metatheorems in Arithmetic

In this section we prove a few metatheorems pertaining to the theories N and PA. They are needed to prove the first incompleteness theorem.

Proposition 4.3.1 *For any formula A of N and any $n \in \mathbb{N}$,*

$$N \vdash A_v[k_0] \to \cdots \to A_v[k_{n-1}] \to v < k_n \to A.$$

Proof. We prove the result by induction on n. For $n = 0$, the result follows since $\neg(v < 0)$ is an axiom of N.

Let the result be true for some n. By the axiom (8) of N,

$$N \vdash v < k_{n+1} \leftrightarrow v < k_n \vee v = k_n.$$

By the equality theorem,

$$N \vdash v = k_n \to (A \leftrightarrow A_v[k_n]).$$

Hence, by the induction hypothesis and the tautology theorem,

$$N \vdash A_v[k_0] \to \cdots \to A_v[k_{n-1}] \to A_v[k_n] \to v < k_{n+1} \to A.$$

□

Proposition 4.3.2 *Let $N \vdash \neg A_v[k_i]$ for all $i < n$ and $N \vdash A_v[k_n]$. Then*

$$N \vdash A \wedge \forall w(w < v \to \neg A_v[w]) \leftrightarrow v = k_n.$$

Proof. Let B denote the formula

$$A \wedge \forall w(w < v \to \neg A_v[w]).$$

By the equality theorem,

$$N \vdash v = k_n \to (B \leftrightarrow B_v[k_n]). \quad (1)$$

By Proposition 4.3.1, we have

$$N \vdash \neg(A_v[w])_w[k_0] \to \cdots \to \neg(A_v[w])_w[k_{n-1}] \to w < k_n \to \neg A_v[w]. \quad (2)$$

Hence, by the hypothesis, the detachment rule, and the generalization rule,

$$N \vdash \forall w(w < k_n \to \neg A_v[w]). \tag{3}$$

Since $N \vdash A_v[k_n]$, by (1) and (3) and the tautology theorem,

$$N \vdash v = k_n \to B. \tag{4}$$

By the substitution theorem, we have

$$N \vdash \forall w(w < v \to \neg A_v[w]) \to (k_n < v \to \neg A_v[k_n]).$$

But $N \vdash A_v[k_n]$. Hence, by the tautology theorem,

$$N \vdash B \to \neg(k_n < v). \tag{5}$$

Since $N \vdash \neg A_v[k_i]$, $i < n$, by Proposition 4.3.1 and the detachment rule,

$$N \vdash v < k_n \to \neg A.$$

Hence,

$$N \vdash B \to \neg(v < k_n). \tag{6}$$

By the axiom (9) of N,

$$N \vdash v < k_n \vee v = k_n \vee k_n < v. \tag{7}$$

Hence, by (4)–(7) and the tautology theorem, we get

$$N \vdash B \leftrightarrow v = k_n.$$

□

Example 4.3.3 Peano arithmetic PA is an extension of N.

This will follow if we show that the axiom (9)

$$x < y \vee x = y \vee y < x$$

of the theory N is a theorem of PA. We show this in three steps.

Step 1: $PA \vdash 0 = y \vee 0 < y$.

Let A be the formula $0 = y \vee 0 < y$. Then $PA \vdash A_y[0]$. By the axiom (8) of N (which is also an axiom of PA),

$$PA \vdash A \leftrightarrow 0 < Sy.$$

Also,

$$PA \vdash 0 < Sy \rightarrow A_y[Sy].$$

Hence,

$$PA \vdash A \rightarrow A_y[Sy].$$

So, by the induction axiom of P,

$$PA \vdash A.$$

Step 2: $PA \vdash x < y \rightarrow Sx < Sy$.

Let B be the formula $x < y \rightarrow Sx < Sy$. By the axiom (7) of N,

$$PA \vdash B_y[0].$$

By the axiom (8) of N,

$$PA \vdash B_y[Sy] \leftrightarrow ((x < y \vee x = y) \rightarrow (Sx < Sy \vee Sx = Sy)).$$

So, by the equality axiom,

$$PA \vdash (x < y \vee x = y) \rightarrow (Sx < Sy \vee Sx = Sy).$$

Hence,

$$PA \vdash B \rightarrow B_y[Sy].$$

By the induction axiom of PA,

$$PA \vdash B.$$

Step 3: Let C denote the formula $x < y \vee x = y \vee y < x$. Since $PA \vdash A$ (Step 1),

$$PA \vdash C_x[0].$$

Since $PA \vdash B$ (Step 2), by the axiom (8) of N,

$$PA \vdash x < y \rightarrow (Sx < y \vee Sx = y)$$

and

$$PA \vdash (y < x \vee y = x) \rightarrow y < Sx.$$

Hence,

$$PA \vdash C \rightarrow C_x[Sx].$$

So, by the induction axiom of P,

$$PA \vdash C.$$

Thus PA is a finite extension of N.

Exercise 4.3.4 Let φ be a formula of PA, x, y distinct, such that y does not occur in φ. Show the following:

(a) $PA \vdash \forall x(\forall y(y < x \to \varphi_x[y]) \to \varphi) \to \forall x\varphi$.
(b) $PA \vdash \exists x\varphi \to \exists x(\varphi \wedge \forall y(y < x \to \neg\varphi_x[y]))$.

Exercise 4.3.5 Let φ be a formula of PA in which no variable other than $v_1, \ldots, v_n$ and w are free, $v_1, \ldots, v_n$, w distinct. Suppose $PA \vdash \exists w\varphi$. Let w' be a new variable and ψ the formula

$$\varphi \wedge \forall w'(w' < w \to \neg\varphi_w[w']).$$

Show the following:

(a) $PA \vdash \exists w\psi$.
(b) $PA \vdash \psi \wedge \psi_w[w''] \to w = w''$.

4.4 Consistency and Completeness

A formula A is said to be **undecidable** in a theory T if neither A nor $\neg A$ is a theorem of T; otherwise, the formula is called **decidable** in T.

It is not reasonable to expect that in a theory all formulas would be decidable. For instance, the formula $v = 0$ of N is undecidable in N. (Why?)

A theory T is called **inconsistent** if every formula of T is a theorem of T. Otherwise, the theory is called **consistent**.

While developing a theory axiomatically, one is naturally confronted with the question of the consistency of the theory. Proving consistency of theories is often a challenging task in mathematics. This is one of the most important topics in axiomatic set theory. Interested readers may see the excellent book of Kenneth Kunen [6].

Lemma 4.4.1 *A theory T is inconsistent if and only if there is a formula A such that both A and $\neg A$ are theorems of T.*

Proof. The necessary part of the result is clear. So, assume that A is such that $\vdash A$ as well as $\vdash \neg A$. Take any formula B. By the expansion axiom and Lemma 3.5.1, $\vdash A \vee B$ and $\vdash \neg A \vee B$. So, by the cut rule, $\vdash B \vee B$. Hence, $\vdash B$ by the contraction rule. □

Remark 4.4.2 Since a proof is finite and in particular uses only a finite number of axioms, a theory is consistent if and only if each of its finitely axiomatized parts is consistent.

Lemma 4.4.3 *If T has a model, then T is consistent.*

Proof. Suppose T has a model M and it is inconsistent. Take a closed formula A. Then, both A and $\neg A$ are theorems of T. Hence, by the validity theorem, both are valid in M. This is a contradiction. □

Exercise 4.4.4 Let T' be an extension of T. If T' is consistent, show that T is also consistent. Assume, moreover, that T' is a conservative extension of T. Show that the converse is also true, i.e., if T is consistent, so is T'.

The following is another useful result.

Proposition 4.4.5 *Let B be the closure of A. Then $T \vdash A$ if and only if $T[\neg B]$ is inconsistent.*

Proof. Suppose $T \vdash A$. Then by the closure theorem, $T \vdash B$, and hence $T[\neg B] \vdash B$. Hence, $T[\neg B]$ is inconsistent.

Now assume that $T[\neg B]$ is inconsistent. Then $T[\neg B] \vdash B$. So, by the deduction theorem, $T \vdash \neg B \to B$. Hence, by the tautology theorem, $T \vdash B$. Thus, $T \vdash A$ by the closure theorem. □

Exercise 4.4.6 (Reduction theorem) Let Γ be a set of formulas of T. Then

$$T[\Gamma] \vdash A$$

if and only if there exist $B_1, \ldots, B_n$, each the closure of a formula in Γ, such that

$$T \vdash B_1 \to \cdots \to B_n \to A.$$

Theorem 4.4.7 (Completeness theorem, first form) *A formula A of T is a theorem of T if and only if it is valid in T.*

The only if part of the result is precisely the validity theorem.

There is an equivalent formulation, the form in which we shall prove it, of this result. The proof still requires some preparation. We postpone the proof to the next chapter.

Theorem 4.4.8 (Completeness theorem, second form) *A theory T is consistent if and only if it has a model.*

We now show that the two forms are equivalent.

Assume the first form of the completeness theorem. Let T be consistent. Let B be a closed formula that is not a theorem of T. So, by the first form, it is not valid in T. This, in particular, gives us a model of T.

Now assume the second form of the completeness theorem. By the closure theorem, without loss of generality we assume that A is closed. By Proposition 4.4.5,

$$T \vdash A \Leftrightarrow T[\neg A] \text{ is inconsistent.}$$

By the second form, $T[\neg A]$ is inconsistent if and only if it has no model. Since A is closed, a model of $T[\neg A]$ is a model of T in which A is not valid. Hence, A is a theorem of T if and only if it is valid in T.

A theory T is called **complete** if it is consistent and if every closed formula is decidable in T.

Remark 4.4.9 The importance of giving a complete set of axioms in the above sense cannot be overemphasized. Let T be a theory with a model M. Let T' be the simple extension of T whose nonlogical axioms are precisely those sentences that are valid in M. We designate this theory by $Th(M)$. Clearly $Th(M)$ is complete. However, we may not be able to mechanically decide whether a sentence is valid in M. This is obviously not a satisfactory situation. A theory for which there is an algorithm to decide whether a formula is an axiom is called an *axiomatized theory.* In an epoch making discovery, Gödel showed that for most of the theories this is impossible. After introducing the notion of an algorithm, we shall briefly study axiomatized theories in Chapter 6.

Exercise 4.4.10 Let T be a complete theory and A, B closed formulas of T. Show that

(a) $T \vdash A \vee B \Longleftrightarrow (T \vdash A$ or $T \vdash B)$.
(b) $T \vdash A \wedge B \Longleftrightarrow (T \vdash A$ and $T \vdash B)$.

We close this section by proving the following theorem, due to Adolf Lindenbaum, which will be used to prove the completeness theorem in the next chapter. This will be proved by Zorn's lemma and Proposition 4.4.5.

Theorem 4.4.11 (Lindenbaum's theorem) *Every consistent theory T admits a simple complete extension.*

Proof. Let $\mathbb{P}$ be the family of subsets Γ of the set of formulas of T such that $T[\Gamma]$ is consistent. Since T is consistent, $\mathbb{P} \neq \emptyset$. $(\emptyset \in \mathbb{P}.)$ Partially order $\mathbb{P}$ by inclusion $\subset$. Let $\mathbb{C}$ be a chain in $\mathbb{P}$ and $\Gamma = \cup\mathbb{C}$. Since every proof is finite, $\Gamma \in \mathbb{P}$. So, by Zorn's lemma, $\mathbb{P}$ has a maximal element, say Δ.

Set $T' = T[\Delta]$. Clearly, T' is a simple consistent extension of T.

We claim that T' is complete. Let A be a closed formula of T that is not a theorem of T'. In particular, $A \notin \Delta$. We must show that $T' \vdash \neg A$. If not, then by Proposition 4.4.5, $T'[A]$ is consistent, contradicting the maximality of Δ. $\square$

5

Completeness Theorem and Model Theory

5.1 Completeness Theorem

In this section we prove the completeness theorem for first-order logic. We shall prove it in its second form (Theorem 4.4.8). The result for countable theories was first proved by Gödel in 1930. The result in its complete generality was first observed by Malcev in 1936. The proof given below is due to Leo Henkin.

Theorem 5.1.1 (Completeness theorem) *Every consistent first-order theory T has a model.*

Since we have only syntactical objects at hand, a model of T has to be built out of these. Since syntactical objects that designate individuals of a model are variable-free terms of the language of T, it seems quite natural to start with these. However, T may have no constants. In that case, we add at least one constant to T and no other nonlogical axiom. By the theorem on constants, the theory T' thus obtained is a conservative extension of T, and so consistent. Further, the restriction of any model of T' to $L(T)$ is a model of T. Thus, without any loss of generality, we assume that T has at least one constant symbol.

Let N be the set of all variable-free terms. Let a and b be variable-free terms such that $T \vdash a = b$. Then interpretations of a and b in any model of T are the same individuals. This leads us to define a binary relation on N as follows:

$$a \sim b \text{ if } T \vdash a = b,$$

where a, b belong to N.

Lemma 5.1.2 *The binary relation $\sim$ is an equivalence relation on N.*

Proof. (i) The relation $\sim$ is reflexive, i.e., for every variable-free term t,

$$T \vdash t = t.$$

This is so by the identity axiom and the substitution rule.

(ii) It is symmetric by the symmetry theorem (Proposition 4.2.15) and the tautology theorem.

(iii) By the equality axiom and the substitution rule,

$$\vdash s = t \to t = u \to s = u.$$

Hence, the relation is transitive by the detachment rule. □

We set M to be the set of $\sim$-equivalence classes. For any $a \in N$, let $[a]$ denote the equivalence class containing a. Since T has constant symbols, M is nonempty. The set M will be the universe of our intended model.

We now define the interpretations of the nonlogical symbols of T in M in a natural way:

$$c_M = [c]$$
$$f_M([a_1], \ldots, [a_n]) = [fa_1 \cdots a_n],$$

and

$$p_M([a_1], \ldots, [a_n]) \text{ if and only if } T \vdash pa_1 \cdots a_n.$$

In the above definitions, c is a constant symbol (so a variable-free term), $a_1, \ldots, a_n$ are variable-free terms, f an n-ary function symbol, and p an n-ary relation symbol. The above functions and relations on M are well-defined by the equality theorem (Proposition 4.2.16).

We have now defined a structure M for the language of T. The structure M is called the **canonical structure** for the language of T. By induction on the length of expressions, the following result is quite routine to prove.

Lemma 5.1.3 *Let a be a variable-free term and A a variable-free open formula of T. Then*

(a) $a_M = [a]$.
(b) $T \vdash A \Longleftrightarrow M \models A$.

Is the canonical structure for $L(T)$ a model of T? A moment's reflection will tell us that T ought to have many constants. For instance, suppose $T \vdash \exists vA$, where v is the only free variable in A. In order that $\exists vA$ be valid in M, there should be a variable-free term t such that $T \vdash A_v[t]$.

In a very special case, we prove that the canonical structure of $L(T)$ is a model of T.

A theory T is called a **Henkin theory** if for every closed formula of the form $\exists x A$ there is a constant symbol, say c, in $L(T)$ such that

$$T \vdash \exists x A \rightarrow A_x[c].$$

An extension of T that is Henkin is called a **Henkin extension** of T.

We have the following theorems.

Theorem 5.1.4 *Every theory T has a conservative Henkin extension T'. In particular, if T is consistent, so is T'.*

We shall give a proof of this later in the section.

So, we fix a conservative Henkin extension T' of T. Since T is consistent, T' is consistent. By Lindenbaum's theorem (Theorem 4.4.11), T' has a complete simple extension T''. Since T'' is a simple extension of a Henkin theory, it is clearly a Henkin theory.

So, the completeness theorem will follow from the following theorem.

Theorem 5.1.5 *If T'' is a complete Henkin theory, then the canonical structure for T'' is a model of T''.*

Proof. By the closure theorem, the result will be proved if we show the following:

For every closed formula A,

$$T'' \vdash A \text{ if and only if } M \models A. \tag{$*$}$$

The proof of $(*)$ proceeds by induction on the number of times the logical symbols $\vee$, $\neg$, and $\exists$ occur in A. This number is called the **height** of A.

By definition of M, $(*)$ holds for all atomic A.

Suppose B is the closed formula $\neg A$ and that $(*)$ holds for A. Let $T'' \vdash B$. Since T'' is consistent, it follows that $T'' \nvdash A$. By the induction hypothesis, it follows that $M \not\models A$. But then $M \models B$. Conversely, suppose $T'' \nvdash B$. Since T'' is complete, this implies that $T'' \vdash A$. By the induction hypothesis, $M \models A$. So $M \not\models B$.

Now suppose $A = B \vee C$, and $(*)$ holds for B and C. We can show that $(*)$ holds for A using similar arguments and Exercise 4.4.10.

Suppose B is the closed formula $\exists x A$ and $(*)$ holds for all formulas of height less than the height of B. Let $T'' \vdash \exists x A$. Since T'' is a Henkin theory, there is a constant symbol c such that

$$T'' \vdash \exists x A \rightarrow A_x[c].$$

By the detachment rule, it follows that

$$T'' \vdash A_x[c].$$

By the induction hypothesis, $M \models A_x[c]$. Let $m = c_M$. Since $c_M = (i_m)_M$, $M \models A_x[i_M]$. Hence $M \models B$. Conversely, suppose $M \models B$. Then there is an $m \in M$ such that $M \models A_x[i_m]$, i_m the name for m in the language L_M. Let $m = [a]$ for some variable-free term a. By Lemma 5.1.3, $a_M = (i_m)_M = m$, i.e., $M \models a = i_m$. Hence, $M \models A_x[a]$. By the induction hypothesis,

$$T'' \vdash A_x[a].$$

Since $A_x[a] \to \exists x A$ is a substitution axiom, by the detachment rule,

$$T'' \vdash B.$$

□

It only remains to prove Theorem 5.1.4. We first define T'. Set $T_0 = T$. Suppose T_n has been defined. We define an extension T_{n+1} of T_n as follows: for each closed formula of T_n of the form $\exists x A$, add a new symbol, say $c_{\exists x A}$, and a new axiom $\exists x A \to A_x[c_{\exists x A}]$. The new constant symbols added to T_{n-1} to define T_n will be called **special constants of level** n; If c is the special constant for $\exists x A$, the axiom $\exists x A \to A_x[c]$ will be called the **special axiom for** c.

Now let T' be the theory whose language is the "union" of the languages of the T_n's, i.e., nonlogical symbols (nonlogical axioms) of T' are precisely those that are nonlogical symbols (respectively nonlogical axioms) of some T_n. Clearly, T' is a Henkin theory.

Claim. *The Henkin extension T' of T is a conservative extension of T.*

Proof of the claim. Our claim will be established if we show that for each n, T_{n+1} is a conservative extension of T_n. To see this, observe the following: Let A be a formula of T and a theorem of T'. Since every proof of A in T' contains only finitely many axioms and finitely many special constants, $T_n \vdash A$ for some n.

Let T_n^c be the extension of T_n obtained by adding all special constants of level $n + 1$ but no new nonlogical axiom. By the theorem on constants, T_n^c is a conservative extension of T_n. Hence, our proof will be complete if we show that T_{n+1} is a conservative extension of T_n^c.

Let A be a formula of T_n^c such that $T_{n+1} \vdash A$. By the reduction theorem, there exist distinct special axioms for special constants of level $n + 1$, $B_1, \ldots, B_k$, such that

$$T_n^c \vdash B_1 \to \cdots \to B_k \to A.$$

Suppose B_1 is the formula $\exists x C \to C_x[c]$, where c is the special constant for $\exists x C$. Since the B_i's are distinct, it follows that c does not occur in $B_2, \ldots, B_k$.

Let y be a variable not occurring in $B_1 \to \cdots \to B_k \to A$. Then, by the theorem on constants,

$$T_n^c \vdash (\exists x C \to C_x[y]) \to B_2 \to \cdots \to B_k \to A.$$

Hence, by the $\exists$-introduction rule,

$$T_n^c \vdash \exists y(\exists x C \to C_x[y]) \to B_2 \to \cdots \to B_k \to A.$$

Now,

$$T_n^c \vdash \exists x C \to \exists y C_x[y]$$

by the variant theorem. Since y does not occur in $\exists x C$, it follows that

$$T_n^c \vdash \exists y(\exists x C \to C_x[y]).$$

By the detachment rule,

$$T_n^c \vdash B_2 \to \cdots \to B_k \to A.$$

Proceeding similarly, in k steps, we show that

$$T_n^c \vdash A.$$

□

The completeness theorem is a very important result in mathematical logic. It shows that our definition of proof is a correct one. Besides, it is quite useful. Instead of giving the tedious syntactical proofs, now one can establish results using the notion of truth. This is often an easier job. Further, arguments no longer depend on logical axioms and rules of inference.

The model of T obtained as above is called the **Henkin model** of T.

Let κ be an infinite cardinal. Recall that a theory T is called a κ-theory if its language has at most κ nonlogical symbols.

Theorem 5.1.6 *Let κ be an infinite cardinal and T a consistent κ-theory. Then there is a model M of T such that $|M| \leq \kappa$.*

Proof. The model M obtained in the proof is of cardinality at most κ. (We invite the reader to prove it.) □

Exercise 5.1.7 Let $L(T')$ be an extension of $L(T)$. Show that T' is an extension of T if and only if the restriction of every model of T' to $L(T)$ is a model of T.

Exercise 5.1.8 Show that two theories are equivalent if and only if they have the same models.

Proposition 5.1.9 *Any finite set of first-order sentences that is valid in the theory T of torsion-free abelian groups is true in some abelian group with torsion. Hence, the theory of torsion-free abelian groups is not finitely axiomatizable.*

Proof. Let $A_1, \ldots, A_n$ be sentences of T that are valid in T. So, these are theorems of T. Since every proof of A_i, $1 \leq i \leq n$, contains only finitely many axioms P_m of the form

$$\forall x(\neg(x = 0) \rightarrow \neg(mx = 0)),$$

there is a natural number k such that A_i has a proof in T without using axioms P_i for $i > k$. Since there are abelian groups such that the order of each of its elements is $\leq k$, the theory T' that is an extension of the theory of abelian groups and whose new axioms are P_i, $i > k$, is consistent. Hence it has a model. The result now follows. □

Exercise 5.1.10 Show the following:

(i) The theory of torsion-free abelian groups is not finitely axiomatizable.
(ii) The theory of divisible abelian groups is not finitely axiomatizable.
(iii) Any finite set of first-order sentences of the theory of fields that are valid in the theory of fields of characteristic zero are valid in fields of characteristic p for all large p.
(iv) The theory of fields of characteristic 0 is not finitely axiomatizable.

5.2 Interpretations in a Theory

In this section we define an interpretation of a theory T in another theory T'. Semantically, this would mean that given any model of T' one can define a model of T. For instance, starting from the Peano axioms, we construct rational numbers and show that they form a field. Thus we can say that field theory has a model in Peano arithmetic.

Let L an L' be first-order languages. An **interpretation** I of L in L' consists of:

(a) a unary predicate symbol U_I of L', called the **universe** of L;
(b) for each constant symbol c of L, a constant symbol c_I of L';
(c) for each n-ary function symbol f of L, an n-ary function symbol f_I of L';
(d) for each n-ary relation symbol p of L other than $=$, an n-ary relation symbol p_I of L'.

Let I be an interpretation of L in L' as above and t a term of L. The term, designated by t_I, of L' obtained from t by replacing each nonlogical symbol u of L by u_I is called the **interpretation of** t **by** I.

An **interpretation** of L in a theory T' is an interpretation I of L in $L(T')$ such that

$$T' \vdash U_I(c_I), \tag{1}$$

$$T' \vdash \exists x U_I x, \tag{2}$$

and

$$T' \vdash U_I x_1 \to \cdots \to U_I x_n \to U_I f_I x_1 \ldots x_n \tag{3}$$

for each n-ary function symbol f of L.

The first condition requires that T' prove that the universe U_I contains c_I; the second requires that T' prove that U_I is nonempty; the third requires that in the theory T', f_I be an n-ary function whose restriction to U_I takes values in U_I. An interpretation I of L in T' may be thought of as a structure of L in T' where the underlying universe is U_I.

Let A be a formula of L. We now proceed to define a formula A^I of L' such that $T' \vdash A^I$ will mean that A is true in the structure U_I. Let A_I be the formula of $L(T')$ obtained from A by replacing each nonlogical symbol u occurring in A by u_I and also replacing each subformula of A of the type $\exists x B$ by $\exists x(U_I x \wedge B_I)$. More precisely, we define A_I by induction on the rank of A. For atomic formulas, we obtain A_I from A by replacing each nonlogical symbol u occurring in A by u_I. If A is $\neg B$ or $B \vee C$, A_I is $\neg B_I$ or $B_I \vee C_I$ respectively. If A is $\exists x B$, A_I is $\exists x(U_I x \wedge B_I)$.

Finally, if $v_0, \ldots, v_{n-1}$ are all the variables free in A (and hence in A_I) in alphabetical order, then A^I is the formula

$$U_I v_0 \to \cdots \to U_I v_{n-1} \to A_I.$$

Note that if A is closed, then A^I is the formula A_I.

An **interpretation** of a theory T in a theory T' is an interpretation I of $L(T)$ in $L(T')$ such that for every nonlogical axiom A of T, $T' \vdash A^I$.

Theorem 5.2.1 *If T has an interpretation in T' and if T' is consistent, then so is T.*

Proof. Let I be an interpretation of T in T' with universe U_I. Since T' is consistent, by the completeness theorem, it has a model M.

Set

$$N = (U_I)_M.$$

By (1) and the validity theorem, N is a nonempty set. For any relation symbol p of T, let p_N be the restriction of $(p_I)_M$ to N. Now take a n-ary function symbol f of T. By (2) and the validity theorem, N is closed under $(f_I)_M$. We define f_N to be the restriction of $(f_I)_M$ to N. Thus, N is a structure for $L(T)$.

Now let A be a nonlogical axiom of T. Then $T' \vdash A^I$. Hence, by the validity theorem, $M \models A^I$. Now it is quite easy to check that $N \models A$. So N is a model of T. $\square$

5.3 Extension by Definitions

In a theory we begin with the minimal number possible of undefined concepts (constant, function, and predicate symbols of the theory). Axioms of the theory state their basic properties. But as the theory develops, more and more concepts are introduced and they are treated as an integral part of the theory. For instance, in number theory, subtraction is not a nonlogical symbol of N. It is defined later. Similarly, in set theory, $\subset$ (inclusion) is a defined concept and not a nonlogical symbol of the language for set theory. Results proved using these concepts are taken as theorems of the original theory. In this section, we show that this is a logically correct process.

Let $\varphi[v_1, \ldots, v_n]$, v_i's distinct, be a formula of T. We form an extension T' of T by adding a new n-ary relation symbol p and adding a new nonlogical axiom

$$pv_1 \cdots v_n \longleftrightarrow \varphi. \tag{1}$$

The formula (1) is called the **defining axiom** for p.

Example 5.3.1 In ZF, if we add a binary relation symbol $\subset$ and a new axiom

$$x \subset y \longleftrightarrow \forall z(z \in x \rightarrow z \in y),$$

we get an extension by definition of ZF in which $\subset$ (**subset**) is a defined concept.

Proposition 5.3.2 *Let $\varphi[v_1, \ldots, v_n]$ be a formula of T and let T' be obtained from T by adding a new n-ary relation symbol p with*

$$pv_1 \cdots v_n \longleftrightarrow \varphi$$

as its defining axiom. Then T' is a conservative extension of T.

Proof. Let A be a formula of T that is a theorem of T'. By the completeness theorem, the proof will be complete if we show that A is valid in T. Let M be a model of T. Interpret p in M as follows: For $a_1, \ldots, a_n \in M$,

$$p(a_1, \ldots, a_n) \Longleftrightarrow M \models \varphi_{v_1,\ldots,v_n}[i_{a_1}, \ldots, i_{a_n}].$$

Thus, we get a model M' of T'. By the validity theorem,

$$M' \models A.$$

But this implies that

$$M \models A.$$

□

Now we consider a similar method of adding a function symbol to a theory. Let $v_0, \ldots, v_{n-1}$ and w, w' be distinct variables and $\varphi[v_0, \ldots, v_{n-1}, w]$ a formula of T. Further, assume that

$$T \vdash \exists w \varphi \qquad (i)$$

and

$$T \vdash (\varphi \wedge \varphi_w[w']) \rightarrow w = w'. \qquad (ii)$$

Informally speaking, conditions (i) and (ii) say that for all $v_0, \ldots, v_{n-1}$, there is a unique w "satisfying" φ. We form T' from T by adding a new n-ary function symbol f and a new nonlogical axiom

$$w = f v_0 \cdots v_{n-1} \longleftrightarrow \varphi. \qquad (iii)$$

The formula (iii) is called the **defining axiom** for f.

Example 5.3.3 In ZF, (after adding $\subset$), consider the following formula $\varphi[x, y]$:

$$\forall z(z \in y \longleftrightarrow z \subset x).$$

Using the power set, comprehension, and extensionality axioms, one shows that the formula φ satisfies (i) and (ii) for $T = ZF$ (in fact the extension of ZF obtained by adding $\subset$). So, one may add a new unary function symbol $\mathcal{P}$ (traditionally called **power set**) and a new nonlogical axiom

$$y = \mathcal{P}(x) \longleftrightarrow \varphi.$$

Remark 5.3.4 Note that if $n = 0$, this method adds a new constant symbol to T. As an example, we can add a constant symbol 0 (called **empty set**) in an extension by definition of ZF. This can be seen as follows: Let $A[y]$ be the formula

$$\forall x \neg(x \in y).$$

Using the set existence, extensionality, and comprehension axioms of ZF, one shows that A satisfies conditions (i) and (ii) for $T = ZF$. One then defines an extension by definition of ZF by adding a new constant symbol 0 and a new axiom

$$y = 0 \leftrightarrow A.$$

Proposition 5.3.5 *Let T' be obtained from T by adding a new n-ary function symbol f with*

$$w = f v_0 \cdots v_{n-1} \longleftrightarrow \varphi$$

as its defining axiom. Then T' is a conservative extension of T.

Proof. Let A be a formula of T that is a theorem of T'. By the completeness theorem, the proof will be complete if we show that A is valid in T. Let M be a model of T. Interpret f in M as follows: For $b, a_0, \ldots, a_{n-1} \in M$,

$$b = f(a_0, \ldots, a_{n-1}) \Longleftrightarrow M \models \varphi_{w,v_0,\ldots,v_{n-1}}[i_b, i_{a_0}, \ldots, i_{a_{n-1}}].$$

Thus, we get a structure M' of $L(T')$ that is an expansion of M. It is easy to check that M' is a model of T'.

By the validity theorem,

$$M' \models A.$$

But this implies that

$$M \models A.$$

□

We say that T' is an **extension by definitions** of T if T' is obtained from T by a finite number of extensions of the two types that we have described.

Theorem 5.3.6 *If T' is an extension by definitions of T, then T' is a conservative extension of T. In particular, T is consistent if and only if T' is consistent.*

An interpretation I of T in T' is called **faithful** if for every formula φ of T, $T' \vdash \varphi^I \Rightarrow T \vdash \varphi$.

Exercise 5.3.7 Suppose T has a faithful interpretation in an extension by definitions of T' and T' is consistent. Show that T is consistent.

Remark 5.3.8 The previous exercise gives a method to prove relative consistency results.

We state some interesting results without proof.

Theorem 5.3.9 *Peano arithmetic PA has an interpretation in an extension by definitions of ZF. [6]*

Theorem 5.3.10 *Each of Peano arithmetic PA and $ZF -$ Infinity has a faithful interpretation in an extension by definitions of the other. In particular, PA is consistent if and only if $ZF -$ Infinity is consistent. [6, Exercise 30, p.149].*

5.4 Compactness Theorem and Applications

In this section we present the compactness theorem for first-order theories. This was first proved for countable theories by Gödel in 1930. In its full generality, it was proved by Malcev. Malcev was also the first who saw the power of this theorem. Using the completeness theorem, it is now trivial to prove.

Theorem 5.4.1 (Compactness theorem for first-order theories) *A theory T has a model if and only if each of its finitely axiomatized parts has a model.*

Proof.

$$\begin{aligned} T \text{ has a model} &\Leftrightarrow T \text{ is consistent} \\ &\Leftrightarrow \text{each finitely axiomatized part of } T \text{ is consistent} \\ &\Leftrightarrow \text{each finitely axiomatized part of } T \text{ has a model.} \end{aligned}$$

□

We now give some applications of the compactness theorem.

Proposition 5.4.2 *Let T be a theory that has arbitrarily large finite models. Then T has an infinite model.*

Proof. Let $\{c_n : n \in \mathbb{N}\}$ be a sequence of distinct symbols not appearing in L. Let T' be the extension of T obtained by adding each c_n as a new constant symbol and for each $m < n$, the formula $\neg(c_n = c_m)$ as a new axiom.

Since T has arbitrarily large finite models, each finite part of T' has a model. Hence, by the compactness theorem, T' has a model. Clearly, any model of T' is infinite and a model of T. □

Remark 5.4.3 This also shows that there is no theory T whose models are precisely the finite sets.

We can use the above idea to build models of arbitrarily large cardinalities.

Theorem 5.4.4 (Tarski) *Let κ be an infinite cardinal. Assume that T has an infinite model M. Then T has a model of cardinality at least κ.*

Proof. Fix a set $\{c_\alpha : \alpha < \kappa\}$ of cardinality κ of distinct symbols not appearing in L. Let L' be the extension of L obtained by adding each c_α as a constant symbol. Set

$$\Gamma = \{c_\alpha \neq c_\beta : \alpha \neq \beta\}.$$

Consider the theory

$$T' = T[\Gamma]$$

with language L'.

We claim that T' is finitely satisfiable. To see this, fix a finite subset Γ' of Γ. Let $c_{\alpha_1}, \ldots, c_{\alpha_k}$ be all the new constants that appear in a formula in Γ'. Since M is infinite, there exist distinct elements $b_1, \ldots, b_k$ of M.

Interpret c_{α_i} by b_i, $1 \leq i \leq k$. Then we get a model of T in which Γ' is satsifiable. Thus, by the compactness theorem, T' has a model. Now note that any model of T' is of cardinality at least κ and a model of T. □

Under the hypothesis of Tarski's theorem, we can say more.

Theorem 5.4.5 (Tarski) *Let κ be an infinite cardinal and T a consistent κ-theory. Assume that T has an infinite model M. Then T has a model of cardinality κ.*

Proof. Let T' be the theory obtained from T as in the proof of Theorem 5.4.4. Note that T' is a consistent κ-theory. By Theorem 5.1.6, T' has a model N of cardinality at most κ. Since any model of T' is of cardinality at least κ, $|N| = \kappa$. Thus, we get a model of T of cardinality κ. □

Corollary 5.4.6 (Löwenheim–Skolem) *Every countable consistent theory has a countable model.*

Remark 5.4.7 ZFC is a countable theory. Assume that ZFC is consistent. Then ZFC has a countable model, say V. However, in ZFC we can prove that the set of real numbers $\mathbb{R}$ is uncountable. How can a countable model contain an uncountable set? This paradox is resolved as follows: no bijection from $\mathbb{N}$ (the set of all natural numbers) onto $\mathbb{R}^V$, the set of all reals in V, is in the model V.

Theorem 5.4.8 (Upward Löwenheim–Skolem theorem) *Let κ be an infinite cardinal, L have at most κ nonlogical symbols, and let N be an infinite structure of L of cardinality at most κ. Then there is a structure M of cardinality κ such that N has an elementary embedding in M.*

Proof. Let N_{el} be the theory whose language is L_N and whose axioms are formulas of the form $\varphi[i_{\overline{a}}]$ ($\overline{a} \in N^n, n \geq 0$) that are valid in N. Since N is a model of N_{el}, the theory N_{el} is consistent. Further, the model N of N_{el} is infinite. Hence, by Theorem 5.4.5, N_{el} has a model M of cardinality κ.

Now define $\alpha : N \to M$ by

$$\alpha(a) = (i_a)_M, \ \ a \in N,$$

i.e., $\alpha(a)$ is the meaning of i_a in M. We claim that $\alpha : N \to M$ is an elementary embedding.

Let a_1, a_2 be distinct elements of N. Then

$$N \models \neg(i_{a_1} = i_{a_2}),$$

i.e., $\neg(i_{a_1} = i_{a_2})$ is an axiom of N_{el}. Hence,

$$M \models \neg(i_{a_1} = i_{a_2}),$$

i.e., $\alpha(a_1) \neq \alpha(a_2)$. Thus, α is an injection.

Since M is a model of N_{el}, α is clearly an elementary embedding of N into M. □

The compactness theorem gives nonstandard models of number theory, real numbers, etc.

The set $\mathbb{N}$ with the usual S, $+$, $\cdot$ and $<$ will be called **standard model** of N and of PA. A formula φ of N or of an extension by definitions of N will be called **true** if it is valid in the standard model $\mathbb{N}$.

Proposition 5.4.9 *There is a model M of the theory N having an element b such that for every natural number n, $(\underline{n})_M < b$.*

Proof. Recall that the set of all natural numbers $\mathbb{N} = \{0, 1, 2, \ldots\}$ is a model of N. Take a new constant symbol c and for each natural number m, let A_m be the formula $\underline{m} < c$. Now consider the theory

$$N' = N[\{A_m : m \in \mathbb{N}\}].$$

Since every finite set of natural numbers has an upper bound in $\mathbb{N}$, it follows that N' is finitely satisfiable. Hence, by the compactness theorem, it has a model M. This model has the required properties with $b = c_M$. □

Exercise 5.4.10 Show that there are models of N of arbitrarily large infinite cardinalities.

Proposition 5.4.11 *There is an ordered field $^*\mathbb{R}$ that is not archimedean.*

Proof. Let T be the theory of ordered fields. Consider the extension T' of T obtained by adding a new constant symbol c and the following new axioms:

$$n1 < c, \quad n \in \mathbb{N}.$$

Since the real line $\mathbb{R}$ is a model of each finitely axiomatized part of T', by the compactness theorem, T' is consistent. Let $^*\mathbb{R}$ be a model of T' and $\overline{1}, b$ the interpretations of 1 and c respectively in this model. Then $n\overline{1} < b$ for all n. Thus $^*\mathbb{R}$ is not archimedean. □

Exercise 5.4.12 Show that there is no first-order theory whose models are precisely archimedean ordered fields.

Exercise 5.4.13 Show that there is no first-order theory whose language has only one nonlogical symbol, namely a binary relation symbol, and whose models are precisely the well-ordered sets.

5.5 Complete Theories

In this section we shall study general properties of complete theories.

Proposition 5.5.1 *Let T be a consistent theory. Then the following statements are equivalent:*

(1) The theory T is complete.
(2) Any two models M and N of T are elementarily equivalent.
(3) For any model M of T, T is equivalent to $Th(M)$.

Proof. We first show that (1) implies (2). Let φ be a sentence in T. If $T \vdash \varphi$, then $M \models \varphi$ as well as $N \models \varphi$. If $T \vdash \neg\varphi$, then $N \not\models \varphi$ and $M \not\models \varphi$. Thus (1) implies (2)

Now assume (2). Any model of $Th(M)$ is clearly a model of T. On the other hand, let N be a model of T. By (2), every theorem of $Th(M)$ is valid in N, i.e., N is a model of $Th(M)$. So, T and $Th(M)$ are equivalent by Proposition 5.5.1.

We now prove that (3) implies (1). Let φ be a closed formula. Exactly one of φ, $\neg\varphi$ is in $Th(M)$. Hence, by (3), one of them is a theorem of T. □

Exercise 5.5.2 Show that there exists a nonarchimedean ordered field elementarily equivalent to the field of reals.

Let κ be an infinite cardinal. A consistent κ-theory T is called κ-**categorical** if any two models of T of cardinality κ are isomorphic.

Our interest in this concept stems from the following result of Robert Vaught.

Theorem 5.5.3 (Vaught) *Let κ be an infinite cardinal and T a consistent κ-theory all of whose models are infinite. If T is κ-categorical, T is complete.*

Proof. Suppose a sentence φ is not decidable in T. By Proposition 4.4.5, the theories $T_1 = T[\varphi]$ and $T_2 = T[\neg\varphi]$ are consistent. Since T has no finite models, both T_1 and T_2 have infinite models. So, by Theorem 5.4.5, T_1 and T_2 have models M_1 and M_2 respectively of cardinality κ. Hence, by the hypothesis of the theorem, they are isomorphic. But φ is valid in M_1 and not in M_2. Thus we have arrived at a contradiction. Hence, T is complete. □

Proposition 5.5.4 *The theory DLO of order-dense, linearly ordered sets with no first and no last elements is complete.*

Proof. By Exercise 2.4.3, DLO is $\aleph_0$-categorical, where $\aleph_0$ is the first infinite cardinal. Clearly all its models are infinite. Therefore by Vaught's theorem (Theorem 5.5.3), DLO is a complete theory. □

5.6 Applications in Algebra

Let $(G, +)$ be a divisible torsion-free abelian group. Then for every $x \in G$ and every $n \geq 1$, there is a unique $y \in G$ such that $ny = x$. We set $y = x/n$. Now for any rational number m/n, we define $\frac{m}{n}x = m(x/n)$. It is easy to check that $(G, +)$ with this definition of scalar product is a vector space over the field of rationals $\mathbb{Q}$.

Proposition 5.6.1 *Any two divisible torsion-free abelian uncountable groups G_1 and G_2 of the same cardinality are isomorphic.*

Proof. We treat both G_1 and G_2 as vector spaces over the field $\mathbb{Q}$ of rationals. Let B_i be a basis of of G_i, $1 \leq i \leq 2$. Since each G_i is uncountable, we have

$$|B_1| = |G_1| = |G_2| = |B_2|.$$

Hence, G_1 and G_2 are isomorphic as vector spaces. In particular, they are isomorphic as groups. □

Hence by Vaught's theorem (Theorem 5.5.3), we have the following result.

Theorem 5.6.2 *The theory of divisible torsion-free abelian groups is complete.*

It is known that two algebraically closed fields are isomorphic if and only if they have the same characteristic and same transcendence degree. Let κ be an uncountable cardinal. It is known that an algebraically closed field is of transcendence degree κ if and only if it is of cardinality κ. [7]. So we have the following theorem.

Theorem 5.6.3 *Let κ be an uncountable cardinal. Then $ACF(0)$ and $ACF(p)$, $p > 1$, are κ-categorical.*

Let $\mathbb{F} = \{a_1, \ldots, a_k\}$ be a finite field and let 1 denote its multiplicative identity. Then the polynomial

$$1 + (x - a_1) \cdot (x - a_2) \cdots (x - a_n)$$

has no zero in $\mathbb{F}$. Hence every algebraically closed field is infinite. Thus all models of ACF are infinite. Now by Vaught's theorem (Theorem 5.5.3), we have the following result.

Theorem 5.6.4 *The theories $ACF(0)$ and $ACF(p)$, $p > 1$, are all complete.*

Theorem 5.6.5 *Let φ be a sentence of the language of the theory of rings with identity. Then the following statements are equivalent:*

(i) φ is valid in the field $\mathbb{C}$ of complex numbers.

(ii) *φ is valid in all algebraically closed fields of characteristic* 0.
(iii) *φ is valid in some algebraically closed field of characteristic* 0.
(iv) *There is an m such that for all prime $p > m$, φ is valid in some algebraically closed field of characteristic p.*
(v) *There is an m such that for all prime $p > m$, φ is valid in all algebraically closed fields of characteristic p.*

Proof. By Proposition 5.5.1, (i) implies (ii) and also (iii) implies (ii). Now assume (ii). Then by the completeness theorem, φ is theorem of $ACF(0)$. Since a proof of φ contains only finitely many nonlogical axioms, there is an m such that for all primes $p > m$, φ is a theorem of $ACF(p)$. Hence (iv) is true by the validity theorem.

The statement (iv) implies (v) because each $ACF(p)$ is complete. We now show that (v) implies (ii). Let $ACF(0) \not\models \varphi$, i.e., φ is not valid in $ACF(0)$. Since $ACF(0)$ is complete, it follows that $\neg\varphi$ is valid in $ACF(0)$. Since (ii) implies (v), it follows that there is an m' such that for all primes $p > m'$, $\neg\varphi$ is valid in all algebraic closed fields of characteristic p. So, (v) is false. □

We close this chapter by giving an application in algebra. We recall a standard result from algebra.

Proposition 5.6.6 *Let p be a prime, F_p the field with p elements, and $\overline{F_p}$ its algebraic closure. Then*

$$\overline{F_p} = \cup_{n \geq 1} F_{p^n}.$$

In particular, every subfield of $\overline{F_p}$ generated by finitely many elements is finite.

Proposition 5.6.7 *Let $p > 1$ be prime and*

$$f_1(X_1, \ldots, X_n), \ldots, f_n(X_1, \ldots, X_n) \in \overline{F_p}[X_1, \ldots, X_n].$$

If the map

$$f = (f_1, \ldots, f_n) : \overline{F_p}^n \to \overline{F_p}^n$$

is injective, it is surjective.

Proof. Suppose f is not surjective. Let $\{a_1, \ldots, a_k\}$ be the set of all coefficients of $f_1, \ldots, f_n$. Let $b_1, \ldots, b_n \in \overline{F_p}$ be such that $(b_1, \ldots, b_n)$ is not in the range of f. Let $\mathbb{K}$ be the subfield of $\overline{F_p}$ generated by $\{a_1, \ldots, a_k, b_1, \ldots, b_n\}$. By Proposition 5.6.6, $\mathbb{K}$ is finite. But now we have an injective map $f : \mathbb{K}^n \to \mathbb{K}^n$ that is not surjective. This is impossible since $\mathbb{K}^n$ is a finite set. □

Theorem 5.6.8 (Ax) *Let $\mathbb{K}$ be an algebraically closed field and*

$$f_1(X_1, \ldots, X_n), \ldots, f_n(X_1, \ldots, X_n) \in \mathbb{K}[X_1, \ldots, X_n].$$

If the map

$$f = (f_1, \ldots, f_n) : \mathbb{K}^n \to \mathbb{K}^n$$

is injective, it is surjective.

Proof. Let each f_i be of degree at most d. It is not hard to see that there is a sentence φ of the language of ring theory saying that if $f_1, \ldots, f_n$ are polynomials of degree at most d and if the map $f = (f_1, \ldots, f_n)$ is injective, then it is surjective.

Let $\mathbb{K}$ be of characteristic p for some prime $p > 1$. Then φ is valid in $\overline{F_p}$ by Proposition 5.6.7. So φ is valid in $\mathbb{K}$ by Theorem 5.6.4 and Proposition 5.5.1.

Now let $\mathbb{K}$ be of characteristic 0. In this case the result follows from Proposition 5.6.7 and Theorem 5.6.5. □

6

Recursive Functions and Arithmetization of Theories

Let us ask the following question: is there an algorithm to decide whether an arbitrary sentence of the theory N is a theorem? Many important mathematical problems are of this type. For instance, the famous **Hilbert's tenth problem** sought for an algorithm to decide whether an arbitrary polynomial equation

$$F(X_1, \ldots, X_n) = 0$$

with integer coefficients (also known as a Diophantine equation) has a solution in rational numbers. Problems of this form are called **decision problems**.

The possibility of nonexistence of an algorithm for a decision problem calls for defining the notion of algorithm precisely. This was considered by several logicians including Herbrand, Church, Kleene, Gödel, and Turing. Several possible definitions were advanced, and quite remarkably, all of them were shown to be equivalent. The notion of algorithm is quite important for the incompleteness theorems. We shall adopt the definition given by Gödel. He introduced a class of functions $f : \mathbb{N}^k \to \mathbb{N}^l$, now called *recursive functions*. These are all the functions that can be computed mechanically. The definition given by Gödel is quite mathematical and helps to prove quite a strong form of incompleteness theorem.

In this chapter we shall study recursive functions. We shall also introduce techniques to show how a general decision problem can be converted into showing whether a partcular function is recursive.

6.1 Recursive Functions and Recursive Predicates

Throughout this and the next section, unless otherwise stated, by a number we shall mean a natural number, by a relation or a predicate we shall mean an n-ary relation on $\mathbb{N}$, $n \geq 1$, and by a function we shall mean a function of the form $f : \mathbb{N}^m \to \mathbb{N}^n$, $m, n \geq 1$. A sequence of numbers $(n_0, \ldots, n_{k-1})$ will usually be denoted by $\overline{n}$. Further, we shall not distinguish between a k-ary relation and a subset of $\mathbb{N}^k$. In fact, in our context, it is more convenient and natural to treat a subset of $\mathbb{N}^k$ as a k-ary relation. Thus, for $P, Q \subset \mathbb{N}^k$,

$$\neg P = P^c = \mathbb{N}^k \setminus P, \quad P \vee Q = P \cup Q, \quad P \wedge Q = P \cap Q,$$

and

$$P \to Q = P^c \cup Q, \quad P \leftrightarrow Q = (P \to Q) \cap (Q \to P).$$

If $Q \subset \mathbb{N}^{k+1}$, we have the following equivalence:

$$\exists m Q(m, \overline{n}) \leftrightarrow (\pi_1 Q)(\overline{n}),$$

where $\pi_1 Q$ denotes the projection of Q to the last $\mathbb{N}^k$ coordinate space.

The **characteristic function** χ_A of $A \subset X$ is defined by

$$\chi_A(x) = \begin{cases} 0 \text{ if } x \in A, \\ 1 \text{ otherwise.} \end{cases}$$

We caution the reader that some authors define $\chi_A(x)$ to be 1 if $x \in A$ and 0 otherwise.

Let P be a k-ary relation. We define a $(k-1)$-ary function by

$$\mu m P(m, \overline{n}) = \begin{cases} 0 & \text{if } \forall m \neg P(m, \overline{n}), \\ \text{first } m \text{ such that } P(m, \overline{n}) \text{ holds} & \text{otherwise,} \end{cases}$$

where $\overline{n} = (n_0, \ldots, n_{k-2})$. In paricular, if k is 1, this defines the following natural number:

$$\mu m P(m) = \begin{cases} 0 & \text{if } \forall m \neg P(m), \\ \text{first } m \text{ such that } P(m) \text{ holds} & \text{otherwise.} \end{cases}$$

The operation μ is called **minimalization**. In the sequel, we shall also need **bounded minimalization** $\mu^<$ and **bounded quantifiers** $\exists^<$, $\forall^<$, $\exists^\leq$, and $\forall^\leq$. We define

$$\mu^{<m} P(m, \overline{n}) \leftrightarrow \mu k[P(k, \overline{n}) \vee k = m],$$

$$\exists^{<m} P(m, \overline{n}) \leftrightarrow \exists k[k < m \wedge P(k, \overline{n})],$$

and

$$\forall^{<m} P(m, \overline{n}) \leftrightarrow \forall k[k < m \rightarrow P(k, \overline{n})].$$

The bounded quantifiers $\exists^{\leq}$ and $\forall^{\leq}$ are similarly defined.

We give below examples of some simple functions.

Successor Function: $S(n) = n + 1$;

Constant Functions: For any $k \geq 1$ and any $p \geq 0$,

$$C_p^k(n_1, \ldots, n_k) \equiv p;$$

Projection Functions: For any $k \geq 1$ and $1 \leq i \leq k$,

$$\pi_i^k(n_1, \ldots, n_k) = n_i.$$

The functions $+$ (addition), $\cdot$ (multiplication), $\chi_<$, and π_i^k, $k \geq 1, 1 \leq i \leq k$, will be called **initial functions**.

Now we fix some constructive schemes for defining a function f from given functions.

Composition: Given $h(n_1, \ldots, n_m)$ and $g_i(l_1, \ldots, l_k)$, $1 \leq i \leq m$, define

$$f(l_1, \ldots, l_k) = h(g_1(l_1, \ldots, l_k), \ldots, g_m(l_1, \ldots, l_k)).$$

Minimalization: Given a function g of $(m + 1)$ variables such that for every $(n_1, \ldots, n_m)$, there is a k such that $g(k, n_1, \ldots, n_m) = 0$, we define f by

$$f(n_1, \ldots, n_m) = \mu k[g(k, n_1, \ldots, n_m) = 0].$$

A function f is called **recursive** if it can be defined by successive applications of composition and minimalization starting with initial functions. More precisely, the set of recursive functions is the smallest collection of functions that contains all initial functions and that is closed under composition and minimalization. A relation R is called **recursive** if its characteristic function χ_R is recursive.

Note that by definition, $<$ is a binary recursive predicate.

It has been accepted that *a function is "computable mechanically" if and only if it is recursive.* The statement in italics is known as **Church's thesis**. Once again we mention that several natural definitions of mechanically computable functions were given. All definitions are shown to be equivalent.

We shall see that many decision problems can be turned into showing whether a function $f : \mathbb{N}^m \rightarrow \mathbb{N}^n$ is computable.

Remark 6.1.1 It is quite easy to see that the sets of recursive functions and recursive predicates are countable. So, there are functions and predicates that are not recursive. However, it is not easy to give examples of such functions and prediactes. This is because nonrecursive functions and nonrecursive predicates are, in some sense, nonconstructive.

Now we proceed systematically to give examples and closure properties of recursive functions and predicates.

If $\pi : \mathbb{N}^k \to \mathbb{N}^n$ is the projection to the first n coordinate spaces and if f is a an n-ary recursive function, so is the k-ary map g defined by

$$g(\overline{u}) = f(\pi(\overline{u})).$$

Lemma 6.1.2 *If $P(\overline{n})$ is a k-ary recursive predicate and if $f_i(\overline{u})$, $1 \leq j \leq k$, are recursive, then the predicate*

$$Q(\overline{u}) \Longleftrightarrow P(f_1(\overline{u}), \ldots, f_k(\overline{u}))$$

is recursive. In particular, if π is a permutation of $\{0, 1, \ldots, n-1\}$ and P an n-ary recursive predicate, then so is the predicate Q defined by

$$Q(l_0, \ldots, l_{n-1}) \leftrightarrow P(l_{\pi(0)}, \ldots, l_{\pi(n-1)}).$$

Proof. Since the set of all recursive functions is closed under composition, the result follows from the following identity:

$$\chi_Q(\overline{u}) = \chi_P(f_1(\overline{u}), \ldots, f_k(\overline{u})).$$

□

The above closure property of the set of all recursive predicates is called closure under **recursive substitutions**.

Lemma 6.1.3 *If $P(m, \overline{n})$ is a recursive predicate such that for every $\overline{n}$, $P(m, \overline{n})$ holds for some m, then*

$$f(\overline{n}) = \mu m P(m, \overline{n})$$

is recursive.

Proof. The result follows from the identity

$$f(\overline{n}) = \mu m[\chi_P(m, \overline{n}) = 0]$$

and the minimalization rule. □

Lemma 6.1.4 *Every constant function C_p^k, $k \geq 1$, $p \geq 0$, is recursive. In particular, $\emptyset$ and each $\mathbb{N}^k$ are recursive.*

Proof. For each k, we prove the result by induction on p. Since

$$C_0^k(\overline{n}) = \mu m[\pi_1^{k+1}(m, \overline{n}) = 0],$$

C_0^k is recursive. Assume that C_p^k is recursive. Now,

$$C_{p+1}^k = \mu m[C_p^k < m].$$

The result follows by the induction hypothesis. □

In what follows, instead of giving complete proofs, we shall define functions and predicates in such a way that it would not be hard to show that they are recursive.

Example 6.1.5 The successor function $S(n) = n + 1$ is recursive. This follows from the identity

$$S(n) = \pi_1^1(n) + C_1^1(n).$$

Example 6.1.6 The binary predicates $\leq$, $>$, and $\geq$ are recursive. This follows from the following equivalences and the closure properties of recursive predicates already proved:

$$m \leq n \Longleftrightarrow m < n + 1,$$

$$m > n \Longleftrightarrow n < m,$$

$$m = n,$$

and

$$m \geq n \Longleftrightarrow m + 1 > n.$$

Proposition 6.1.7 *If P and Q are n-ary recursive predicates, so are $\neg P$ and $P \vee Q$. It follows that the predicates $P \wedge Q$, $P \rightarrow Q$, and $P \leftrightarrow Q$ are also recursive if P and Q are. In particular each finite subset of $\mathbb{N}^k$, $k \geq 1$, is recursive.*

Proof. The result follows from the following identity:

$$\chi_{\neg P}(\overline{p}) = \chi_{<}(0, \chi_P(\overline{p})) \text{ and } \chi_{P \vee Q}(\overline{p}) = \chi_P(\overline{p}) \cdot \chi_Q(\overline{p}),$$

where $\overline{p} = (p_1, \ldots, p_n)$. □

Example 6.1.8 We define $m \dot{-} n$ as follows:

$$m \dot{-} n = \begin{cases} m - n \text{ if } m \geq n, \\ 0 \qquad \text{otherwise.} \end{cases}$$

The function $m \dot{-} n$ is recursive. To see this, note the following identity:

$$m \dot{-} n = \mu k[n + k = m \lor m < n].$$

Exercise 6.1.9 Show that the following functions are recursive:

(i) $|m - n|$.
(ii) $\min(m, n)$.
(iii) $\max(m, n)$.
(iv)

$$\alpha(n) = \begin{cases} 1 \text{ if } n = 0, \\ 0 \text{ if } n > 0. \end{cases}$$

(v)

$$sg(n) = \begin{cases} 0 \text{ if } n = 0, \\ 1 \text{ if } n > 0. \end{cases}$$

Exercise 6.1.10 Show that the set of all recursive functions is closed under bounded minimalization; the set of all recursive predicates is closed under bounded quantifiers.

Remark 6.1.11 In the next chapter, we shall show that the set of all recursive predicates is not closed under existential and universal quantifiers.

Exercise 6.1.12 Let $A_1, \ldots, A_m$ be pairwise disjoint recursive subsets of $\mathbb{N}^k$ whose union is $\mathbb{N}^k$. Suppose $f_1, \ldots, f_m$ are k-ary recursive functions. Define $g : \mathbb{N}^k \to \mathbb{N}$ by

$$g(\overline{a}) = \begin{cases} f_1(\overline{a}) & \text{if } \overline{a} \in A_1, \\ \vdots \\ f_m(\overline{a}) & \text{if } \overline{a} \in A_m. \end{cases}$$

Show that g is recursive.

We now proceed to show that we can effectively code a finite sequence of numbers, a finite sequence of finite sequences of numbers, etc. by numbers. This remarkable idea is due to Gödel, and he turned it into a powerful tool for proving his incompleteness theorems.

Exercise 6.1.13 The following predicates are recursive:

(i) **(Divisibility)** $m|n \leftrightarrow \exists^{\leq n} k[m \cdot k = n]$.
(ii) **(Prime)** $\text{Prime}(p) \leftrightarrow p$ is a prime.
(iii) **(Relatively prime)** $RP(m, n) \leftrightarrow m \neq 0 \land n \neq 0 \land \forall p \leq m[(\text{Prime}(p) \land p|m) \to \neg p|n]$.

For an ordered pair (m, n) of natural numbers, we define

$$OP(m, n) = (m + n) \cdot (m + n + 1) + n + 1.$$

Clearly, $OP(m, n)$ is recursive.

Lemma 6.1.14 *The function OP is one-to-one.*

Proof. Let $OP(m, n) = OP(m', n')$. We first show that $m + n = m' + n'$. Suppose not. Without any loss of generality, we assume that $m + n < m' + n'$. Now

$$OP(m, n) \leq (m + n + 1)^2 \leq (m' + n')^2 < OP(m', n').$$

This is a contradiction.

Since $m + n = m' + n'$ and $OP(m, n) = OP(m', n')$, from the definition of OP, it follows that $n = n'$. This, in turn, implies that $m = m'$ too. □

We shall need the following two simple lemmas from number theory.

Lemma 6.1.15 *Let $m_1, \ldots, m_k$ and $n_1, \ldots, n_l$ be sequences of numbers such that $\forall i \forall j[RP(m_i, n_j)]$. Then there is a number x such that $\forall i[m_i|x]$ and $\forall j[RP(x, n_j)]$.*

Proof. Take x to be the product of all the m_i's. If a prime p divides x, it divides some m_i. Hence, $\neg p|n_j$ for all j. The result follows. □

Lemma 6.1.16 $\forall m, n \forall j[m|n \rightarrow RP(1 + (j + m)n, 1 + jn)]$.

Proof. Let p be a prime number such that $p|1+(j+m)n$ as well as $p|1+jn$. Then $p|mn$. Since p is a prime, either $p|m$ or $p|n$. If $p|m$, then $p|n$ also because $m|n$. Hence, $p|n$. But then $\neg(p|1 + jn)$. This contradiction proves our result. □

The following result is due to Gödel.

Theorem 6.1.17 *There is a 2-ary function $\beta(n, i)$ satisfying the following properties:*

(a) β is recursive.
(b) $\beta(0, i) = 0$ for all i.
(c) $n \neq 0 \rightarrow \beta(n, i) < n$ for all i.
(d) For every finite sequence $\overline{n} = (n_0, \ldots, n_{k-1})$ of positive length, there is an n such that

$$\forall i < k[\beta(n, i) = n_i].$$

Proof. Define

$$\beta(n, i) = \mu^{<n} x \exists^{<n} y \exists^{<n} z[n = OP(y, z) \wedge (1 + (OP(x, i) + 1) \cdot z)|y],$$

i.e., $\beta(n, i)$ is the first natural number $x < n$ for which there exist $y, z < n$ satisfying $n = OP(y, z)$ and $y|1 + (OP(x, i) + 1) \cdot z$. If such an x does not exist, $\beta(n, i) = n \dot{-} 1$.

Properties (a)–(c) of β follow from the definition. To prove (d), take a finite sequence $\overline{n} = (n_0, \ldots, n_{k-1})$ of positive length. Let

$$u = \max\{OP(n_i, i) + 1 : i < k\},$$

and z the product of all nonzero numbers less than u. Then, by Lemma 6.1.16,

$$j < l < u \Longrightarrow RP(1 + jz, 1 + lz).$$

For $i < k$, set

$$m_i = 1 + (OP(n_i, i) + 1)z.$$

Let $\{l_j\}$ be an enumeration of all numbers of the form $1 + vz$, where $0 < v < u$ and $v \neq OP(n_i, i) + 1$ for all $i < k$. Then, by Lemma 6.1.15, there is a number y such that

$$\forall j < u[1 + jz|y \Longleftrightarrow \exists i(j = OP(n_i, i) + 1)].$$

Set $n = OP(y, z)$.

It remains to show that $\beta(n, i) = n_i$ for all i. This will follow if we show that n_i is the smallest number x such that

$$1 + (OP(x, i) + 1)z|y.$$

Note that this will follow if for all $x < n_i$, $OP(x, i) < u$ and $OP(x, i) \neq OP(n_j, j)$ for all j. This can easily be seen using the fact that OP is one-to-one (Lemma 6.1.14). □

The function β defined above is called **Gödel's β-function**.

For each $n \geq 1$ and each finite sequence $(k_1, \ldots, k_n)$, we define

$$\langle k_1, \ldots, k_n \rangle = \mu m[\beta(m, 0) = n \wedge \beta(m, 1) = k_1 \wedge \cdots \wedge \beta(m, n) = k_n].$$

We shall call such a number a **sequence number**.

The empty sequence of natural numbers will be denoted by $\langle\ \rangle$, and its sequence number equals 0 by definition.

We now introduce some relations and functions of sequence numbers: seq will denote the set of all sequence numbers, $lh(n) = \beta(n, 0)$ (the length of the sequence coded by n), $(n)_i = \beta(n, i + 1)$ (the ith element of the sequence coded by n), and concatenation

$$\langle m_0, \ldots, m_{l \dot{-} 1} \rangle * \langle n_1, \ldots, n_{k \dot{-} 1} \rangle = \langle m_0, \ldots, m_{l \dot{-} 1}, n_0, \ldots, n_{k \dot{-} 1} \rangle.$$

Proposition 6.1.18 *The following functions and predicates are recursive:* $n \longrightarrow lh(n)$, $(n, i) \longrightarrow (n)_i$, seq, *and the concatenation* $(m, n) \rightarrow m * n$.

Proof. Since $lh(n) = \beta(n, 0)$ and $(n)_i = \beta(n, i+1)$ and since β is recursive, the functions $lh(n)$ and $(n)_i$ are recursive. That seq is recursive follows from the following equivalence:

$$\text{seq}(n) \iff \exists^{<n} u_0, \ldots, u_{lh(n)-1} \forall 0 \leq i < lh(n)[\beta(n, i) = u_i].$$

Let $m = \langle m_0, \ldots, m_{lh(m)-1} \rangle$ and $n = \langle n_0, \ldots, n_{lh(n)-1} \rangle$. Then

$$\begin{aligned} m * n = \mu u[&seq(u) \wedge lh(u) = lh(m) + lh(n) \\ &\wedge \forall i < lh(m)((u)_i = m_j) \wedge \forall j < lh(n)((u)_{l+j} = n_j)]. \end{aligned}$$

Hence the concatenation function $*$ is recursive. □

We introduce yet another operation on recursive functions.

Primitive Recursion: Given an m-ary function g and an $(m+2)$-ary function h, we define an $(m+1)$-ary function f by

$$\begin{aligned} f(0, \overline{n}) &= g(\overline{n}), \\ f(k+1, \overline{n}) &= h(f(k, \overline{n}), k, \overline{n}). \end{aligned}$$

The scheme of primitive recursion is a general form of definition of functions by induction. It should be noted that m may be 0 and that a 0-ary function is nothing but a constant. Thus given a natural number p and a 2-ary function h, this procedure defines a sequence $\{x_k\}$ by induction: set $x_0 = p$ and $x_{k+1} = h(x_k, k)$. Intuitively it should be obvious that if g and h are "computable," so is f.

Proposition 6.1.19 *If g is an m-ary and h an $(m+2)$-ary recursive function and if f is defined by primitive recursion as above, then f is recursive.*

Proof. The function $f : \mathbb{N}^{m+1} \to \mathbb{N}$ is defined by

$$\begin{aligned} f(0, \overline{n}) &= g(\overline{n}), \\ f(k+1, \overline{n}) &= h(f(k, \overline{n}), k, \overline{n}). \end{aligned}$$

We first define a function $F : \mathbb{N}^{m+1} \to \mathbb{N}$ as follows:

$$\begin{aligned} F(p, \overline{n}) = \mu k[&lh(k) = p + 1 \wedge (k)_0 = g(\overline{n}) \\ &\wedge \forall i < p((k)_{i+1} = h((k)_i, i, \overline{n})]. \end{aligned}$$

By the closure properties of recursive functions and recursive predicates, F is recursive. Now note that for all $p \geq 0$,

$$f(p, \overline{n}) = (F(p, \overline{n}))_p.$$

The result can now easily be seen. □

Exercise 6.1.20 Let $\mathcal{R}$ be the smallest set of functions that contains the successor function S, constant functions C_k^n, and the projection maps π_i^n, and that is closed under composition, minimalization, and primitive recursion. Show that a function is recursive if and only if it belongs to $\mathcal{R}$.

The previous exercise gives a more traditional definition of recursive functions. Further, coding a sequence, sequence of sequences, etc. is achieved more easily with this definition. However, our definition of recursive functions is chosen to give the best-known form of Gödel's theorem.

Example 6.1.21 The exponentiation function m^n inductively defined by

$$\begin{aligned} m^0 &= 1, \\ m^{n+1} &= m^n \cdot m, \end{aligned}$$

is recursive.

Exercise 6.1.22 The function $n!$ is recursive.

Exercise 6.1.23 Show that the **predecessor function** $p(n)$ defined below is recursive:

$$\begin{aligned} p(0) &= 0, \\ p(n+1) &= n. \end{aligned}$$

Exercise 6.1.24 Show that the functions $\max\{m, n\}$, $\min\{m, n\}$, and for any $k > 1$, $\max\{n_1, \ldots, n_k\}$ and $\min\{n_1, \ldots, n_k\}$ are recursive.

Exercise 6.1.25 Let $2 = p_0, p_1, p_2, \ldots$ be the increasing enumeration of all prime numbers. Show that $n \to p_n$ is recursive.

Exercise 6.1.26 (Closure under complete recursion) Let $f(m, \overline{n})$ be recursive and $g(m, \overline{n})$ be defined by the equation

$$g(m, \overline{n}) = f(\langle g(0, \overline{n}), \ldots, g(m-1, \overline{n}), \overline{n}).$$

Show that g is recursive. (What is $g(0, \overline{n}?)$

Proposition 6.1.27 *A function $f : \mathbb{N}^k \to \mathbb{N}$ is recursive if and only if its graph $gr(f)$ is recursive, where for any $\overline{n} \in \mathbb{N}^k$,*

$$(\overline{n}, m) \in gr(f) \iff f(\overline{n}) = m.$$

Proof. If f is recursive, $gr(f)$ is recursive because $=$ is recursive and the set of all recursive predicates is closed under recursive substitutions. Conversely, if $gr(f)$ is recursive, f is recursive because of the following identity:

$$f(\overline{n}) = \mu m[(\overline{n}, m) \in gr(f)].$$

□

We prove the following result using the well-known diagonal argument of Cantor. Gödel uses it beautifully to prove the first incompleteness theorem.

Proposition 6.1.28 *There is no recursive set $U \subset \mathbb{N} \times \mathbb{N}$ such that for every recursive set $A \subset \mathbb{N}$ there is an $n \in \mathbb{N}$ satisfying*

$$\forall m[m \in A \iff (n, m) \in U],$$

i.e., there is no recursive set $U \subset \mathbb{N} \times \mathbb{N}$ whose vertical sections exactly list all recursive subsets of $\mathbb{N}$.

Proof. Suppose such a recursive set U exists. Define

$$A^* = \{m \in \mathbb{N} : (m, m) \notin U\}.$$

Since the predicate U^c is recursive and since the set of all recursive predicates is closed under recursive substitutions, A^* is recursive. So, by the hypothesis, there is an $n^* \in \mathbb{N}$ such that

$$\forall m[m \in A^* \iff (n^*, m) \in U]. \qquad (*)$$

If $n^* \in A^*$, $(n^*, n^*) \in U$ by $(*)$. But then $n^* \notin A^*$ by the definition of A^*. On the other hand, if $n^* \notin A^*$, $(n^*, n^*) \in U^c$ by $(*)$. But then $n^* \in A^*$ by the definition of A^*. We have arrived at a contradiction. □

For the next exercise, recall the definition of standard model and true formulas of N.

Exercise 6.1.29 Recall that for any $n \in \mathbb{N}$, k_n denotes the term

$$\underbrace{S \cdots S}_{n \text{ times}} 0$$

of N. Let $\varphi[x_1, \ldots, x_p]$ be an open formula of N. Show that the predicate

$$\{\overline{n} \in \mathbb{N}^p : \varphi_{x_1, \ldots, x_p}[k_{n_1}, \ldots, k_{n_p}] \text{ is true}\}$$

is recursive.

6.2 Semirecursive Predicates

A nonempty subset of $\mathbb{N}^k$ is called **semirecursive** or **recursively enumerable (r.e.**, for short) if it is the projection to the last k coordinate space of a $(k+1)$-ary recursive predicate, i.e., there is a recursive $Q \subset \mathbb{N}^{k+1}$ such that for every $\overline{n} \in \mathbb{N}^k$,

$$P(\overline{n}) \iff \exists m[Q(m, \overline{n})].$$

Proposition 6.2.1 *Every recursive predicate is semirecursive.*

Proof. Let P be a k-ary recursive predicate. Define $Q \subset \mathbb{N}^{k+1}$ by

$$Q(m, \overline{n}) \iff P(\overline{n}).$$

Then Q is recursive and $P(\overline{n}) \iff \exists m Q(m, \overline{n})$. Hence P is semirecursive. □

Proposition 6.2.2 *The set of all semirecursive predicates is closed under* $\vee$, $\wedge$, *projections, bounded universal quantifiers, and recursive substitutions.*

Proof. Let P and Q be k-ary semirecursive predicates. Fix $(k+1)$-ary recursive predicates P' and Q' such that for all $\overline{n} \in \mathbb{N}^k$,

$$P(\overline{n}) \iff \exists m P'(m, \overline{n}) \text{ and } Q(\overline{n}) \iff \exists m Q'(m, \overline{n}).$$

Let f_i, $1 \leq i \leq k$, be l-ary recursive functions. For any $\overline{n} \in \mathbb{N}^k$, $\overline{m} \in \mathbb{N}^l$ note the following:

$$(P \vee Q)(\overline{n}) \iff \exists m[(P' \vee Q')(m, \overline{n})],$$

$$P(f_1(\overline{m}), \ldots, f_k(\overline{m})) \iff \exists r P'(r, f_1(\overline{m}), \ldots, f_k(\overline{m})),$$

and

$$(P \wedge Q)(\overline{n}) \iff \exists m[P'((m)_0, \overline{n}) \wedge Q'((m)_1, \overline{n})].$$

These show that the set of all semirecursive predicates is closed under $\vee$, $\wedge$, and recursive substitutions.

We use Gödel's coding functions to show other closure properties also. Let P be a $(k+2)$-ary recursive predicate and let Q be defined by

$$Q(\overline{n}) \iff \exists l \exists m P(l, m, \overline{n}).$$

Then

$$Q(\overline{n}) \iff \exists m P((m)_0, (m)_1, \overline{n}).$$

This shows that the set of all semirecursive predicates is closed under the existential quantifier.

Let P be a $(k+1)$-ary semirecursive predicate and let Q be defined by

$$Q(m, \overline{n}) \iff \forall^{<m} p P(p, \overline{n}).$$

Let $R \subset \mathbb{N} \times \mathbb{N}^{k+1}$ be a recursive predicate such that

$$\forall p \forall \overline{n} \in \mathbb{N}^k [P(p, \overline{n}) \iff \exists q R(q, p, \overline{n})].$$

Then

$$Q(m, \overline{n}) \iff \exists q[\text{seq}(q) \wedge lh(q) = m \wedge \forall^{<m} p R((q)_p, p, \overline{n})].$$

This shows that the set of all semirecursive predicates is closed under the bounded universal quantifier $\forall^<$. The result is now easily seen. □

Exercise 6.2.3 A nonempty subset P of $\mathbb{N}$ is semirecursive if and only if it is the range of a unary recursive function.

The following is an important result in recursive function theory.

Theorem 6.2.4 (Kleene) *A predicate P is recursive if and only if both P and $\neg P$ are semirecursive.*

Proof. If P is recursive, so is $\neg P$. By Proposition 6.2.1, both P and $\neg P$ are semirecursive.

Now let P be k-ary and both P and $\neg P$ be semirecursive. Choose $(k+1)$-ary recursive predicates Q and R such that for all $\overline{n} \in \mathbb{N}^k$,

$$P(\overline{n}) \iff \exists m Q(m, \overline{n}) \text{ and } \neg P(\overline{n}) \iff \exists m R(m, \overline{n}).$$

Then $S = Q \vee R$ is recursive. Note that

$$\forall \overline{n} \in \mathbb{N}^k \exists m S(m, \overline{n}).$$

We define

$$s(\overline{n}) = \mu m S(m, \overline{n}).$$

The function s is recursive. Further,

$$P(\overline{n}) \iff Q(s(\overline{n}), \overline{n}).$$

This shows that P is recursive. □

Remark 6.2.5 Let $P \subset \mathbb{N}^k$ be semirecursive and let there exist a $(k+1)$-ary recursive predicate Q such that for all $\overline{n}$,

$$P(\overline{n}) \iff \forall m Q(\overline{n}, m).$$

Then P is recursive. By Theorem 6.2.4, our assertion will be proved if we show that $\neg P$ is semirecursive. This follows from the following equivalence:

$$\neg P(\overline{n}) \iff \exists m \neg Q(\overline{n}, m).$$

Proposition 6.2.6 *Let $f : \mathbb{N}^k \to \mathbb{N}$ be any function. Then the following statements are equivalent:*

(*i*) *The function f is recursive.*
(*ii*) *The graph of f, $gr(f)$, is recursive.*
(*iii*) *The graph of f is semirecursive.*

Proof. By Proposition 6.1.27 and Theorem 6.2.4, we only need to show that if $gr(f)$ is semirecursive, $\neg gr(f)$ is semirecursive. This follows from the following equivalence:

$$f(\overline{n}) \neq m \Longleftrightarrow \exists l[m \neq l \wedge f(\overline{n}) = l].$$

□

A function $f = (f_1, \ldots, f_p) : \mathbb{N}^k \to \mathbb{N}^p$ is called **recursive** if each f_i, $1 \leq i \leq p$, is recursive.

Exercise 6.2.7 Let $f : \mathbb{N}^k \to \mathbb{N}^p$ be any function. Then the following statements are equivalent:

(i) The function f is recursive.
(ii) The graph of f, $gr(f)$, is recursive.
(iii) The graph of f is semirecursive.

6.3 Arithmetization of Theories

The next idea, arithmetization of theories, is a beautiful idea due to Gödel. It represents syntactical objects, e.g., symbols, terms, formulas, proofs, of a theory by natural numbers. Consequently, statements about syntactical objects are expressed in terms of numbers. Its importance and beauty cannot be overemphasized. It has the potential of converting a metamathematical statement into a number-theoretic statement. Thus, the problem of whether a metamathematical statement is true is translated into a number-theoretic problem. This idea also plays a significant role in the theory of computation. The same idea is now used to convert many questions concerning algorithms into proving whether a number-theoretic function or relation is recursive. To elaborate a bit more, one can code each algorithm by an integer, or one can translate questions about algorithms into number-theoretic problems.

Throughout this section, unless otherwise stated, T will denote a fixed first-order theory. To simplify the matter, we assume that T is finite and its nonlogical symbols are enumerated in some order.

In the first step we assign a **symbol number** to each symbol of $L(T)$.

Set $SN(x_i) = 2i$, $i \geq 0$; $SN(\neg) = 1$; $SN(\vee) = 3$; $SN(\exists) = 5$; $SN(=) = 7$; if α is the ith nonlogical symbol, we set $SN(\alpha) = 7 + 2i$.

Note that a number n is the symbol number of a variable if and only if it is even, i.e., $2|n$. Hence, the predicate

$$\text{vble}(n) \Longleftrightarrow n \text{ is the symbol number of a variable}$$

is recursive. Since every finite set is recursive, the predicate

$$\text{sn}(n) \Longleftrightarrow n \text{ is a symbol number}$$

is easily seen to be recursive. In other words, there is an algorithm to decide whether an integer is a symbol number. Further, intuitively it is easy to see that there is an algorithm such that given a symbol number n, the algorithm recovers the symbol whose symbol number is n. Let func_0 denote the set of symbol numbers of all constant symbols. We also define

$$\mathrm{pred}(n) \iff n \text{ is the symbol number of a predicate symbol}$$

and

$$\mathrm{func}(n) \iff n \text{ is the symbol number of a function symbol.}$$

Since T is finite, these predicates are recursive.

Let t be a term and A a formula of T. We now define the **Gödel number** $\lceil t \rceil$ and $\lceil A \rceil$ of t and A respectively by induction on the rank of t and A.

If t is a variable or a constant, set

$$\lceil t \rceil = \langle SN(t) \rangle.$$

If f is an n-ary function and $t_1, \ldots, t_n$ are terms whose Gödel numbers have been defined, we set

$$\lceil f t_1 \ldots t_n \rceil = \langle SN(f), \lceil t_1 \rceil, \ldots, \lceil t_n \rceil \rangle.$$

We define

$$\mathrm{term}(n) \iff n \text{ is the Gödel number of a term.}$$

Proposition 6.3.1 *The predicate* term *is recursive.*

Proof. Note that for any n

$$\begin{aligned}\mathrm{term}(n) \iff \mathrm{seq}(n) \wedge [(lh(n) = 1 \wedge \mathrm{func}_0((n)_0)) \vee [lh(n) > 1 \wedge \\ \mathrm{func}((n)_0) \wedge \forall 0 < i < lh(n)(\mathrm{term}((n)_i))]].\end{aligned}$$

Using closure under complete recursion (Exercise 6.1.26), it can now be seen that *term* is recursive. We leave the details as an exercise for the reader. □

We recall that the term

$$\underbrace{S \cdots S}_{n \text{ times}} 0$$

of N is denoted by k_n and that these terms are called numerals.

Lemma 6.3.2 *The map*

$$\mathrm{num}(n) = \lceil k_n \rceil, \quad n \in \mathbb{N},$$

is recursive.

Proof. This follows from the following identities

$$\begin{aligned} \text{num}(0) &= \langle SN(0) \rangle \\ \text{num}(n+1) &= \langle SN(S), \text{num}(n) \rangle. \end{aligned}$$

□

If A is an atomic formula $pt_1 \cdots t_n$, define

$$\ulcorner A \urcorner = \langle SN(p), \ulcorner t_1 \urcorner, \ldots, \ulcorner t_n \urcorner \rangle.$$

Proposition 6.3.3 *The predicate*

$$\text{aform}(n) \Leftrightarrow n \textit{ is the Gödel number of an atomic formula,}$$

is recursive.

Proof. We have

$$\text{aform}(n) \Longleftrightarrow \text{seq}(n) \wedge lh(n) > 1 \wedge \text{pred}((n)_0) \wedge \forall 0 < i < lh(n)[\text{term}((n)_i)].$$

□

If A is $\neg B$, and if $\ulcorner B \urcorner$ has been defined,

$$\ulcorner A \urcorner = \langle SN(\neg), \ulcorner B \urcorner \rangle.$$

If A is $\vee BC$,

$$\ulcorner A \urcorner = \langle SN(\vee), \ulcorner B \urcorner, \ulcorner C \urcorner \rangle.$$

If A is $\exists x B$,

$$\ulcorner A \urcorner = \langle SN(\exists), \ulcorner x \urcorner, \ulcorner B \urcorner \rangle.$$

Proposition 6.3.4 *The predicate*

$$\text{form}(n) \Longleftrightarrow n \textit{ is the Gödel number of a formula}$$

is recursive.

Proof. We define the following predicates:

$$A_1(n) \Longleftrightarrow [\text{seq}(n) \wedge lh(n) = 2 \wedge (n)_0 = SN(\neg) \wedge \text{form}((n)_1)],$$

$$A_2(n) \Longleftrightarrow [\text{seq}(n) \wedge lh(n) = 3 \wedge (n)_0 = SN(\vee) \wedge \text{form}((n)_1) \wedge \text{form}((n)_2)],$$

and

$$A_3(n) \Longleftrightarrow [\text{seq}(n) \wedge lh(n) = 3 \wedge (n)_0 = SN(\exists) \wedge \text{vble}((n)_1) \wedge \text{form}((n)_2)].$$

By closure under complete recursion (Exercise 6.1.26), it is not hard to show that these predicates are recursive. Now note that

$$\text{form}(n) \iff \text{aform}(n) \vee A_1(n) \vee A_2(n) \vee A_3(n).$$

It follows that the predicate form(n) is resursive. □

Now we systematically proceed and show that many metamathematical statements (statements about the theory itself or statements about syntactical objects such as formulas and proofs) can be turned into number-theoretic statements. Many of the functions and predicates thus defined are recursive or semirecursive. However, we shall not verify this in full detail. Interested readers should complete the proofs as an exercise.

Proposition 6.3.5 *There is a recursive function* sub(l, m, n) *such that if* l *is the Gödel number of a term* t *or a formula* A*, if* m *is the Gödel number of a variable* v*, and if* n *is the Gödel number of a term* s*, then* sub(l, m, n) *is the Gödel number of* $t_v[s]$ *or* $A_v[s]$ *respectively.*

Proof. Define

$$\text{sub}_1(l, m, n) = \begin{cases} n \text{ if vble}(l), \wedge l = m, \\ l \text{ otherwise}, \end{cases}$$

$$\text{sub}_2(l, m, n) = \langle (l)_0, \text{sub}_2((l)_1, m, n), \ldots, \text{sub}_2((l)_{lh(l) \dot{-} 1}, m, n) \rangle,$$

$$\text{sub}_3(l, m, n) = \begin{cases} \langle (l)_0, \text{sub}_3((l)_1, m, n) \rangle & \text{if seq}(l) \wedge lh(l) = 2, \\ \langle (l)_0, \text{sub}_3((l)_1, m, n), \text{sub}_3((l)_2, m, n) \rangle & \\ & \text{if seq}(l) \wedge lh(l) = 3 \\ & \wedge (l)_0 \neq SN(\exists), \\ \langle (l)_0, (l)_1, \text{sub}_3((l)_2, m, n) \rangle & \text{if seq}(l) \wedge lh(l) = 3 \\ & \wedge (l)_0 = SN(\exists) \\ & \wedge (l)_1 \neq m, \\ l & \text{otherwise}. \end{cases}$$

Now define

$$\text{sub}(l, m, n) = \begin{cases} \text{sub}_1(l, m, n) \text{ if vble}(l) \vee \text{func}_0(l), \\ \text{sub}_2(l, m, n) \text{ if pred}((l)_0) \vee \text{func}((l)_0), \\ \text{sub}_3(l, m, n) \text{ if form}(l) \wedge \neg\text{aform}(l), \\ l \qquad\qquad\quad\; \text{otherwise}. \end{cases}$$

Then sub(l, m, n) is a recursive function with the desired properties. □

Exercise 6.3.6 For each $n \geq 1$, show that there is a recursive function sb : $\mathbb{N} \times \mathbb{N}^n \to \mathbb{N}$ such that whenever $m = \ulcorner E \urcorner$, E a term or a formula of the theory N,

$$\mathrm{sb}(m, b_0, \ldots, b_{n-1}) = \lceil E_{x_0,\ldots,x_{n-1}}[k_{b_0}, \ldots, k_{b_{n-1}}] \rceil ,$$

where $x_0, x_1, \ldots, x_{n-1}$ are the first n variables in alphabetical order.

Exercise 6.3.7 For each $m, n \geq 1$, show that there is a recursive function $s_n^m : \mathbb{N}\times\mathbb{N}^m \to \mathbb{N}$ such that whenever $p = \lceil A \rceil$, A a formula of the theory N,

$$s_n^m(p, b_{m+1}, \ldots, b_{m+n}) = \lceil A_{x_{m+1},\ldots,x_{m+n}}[k_{b_{m+1}}, \ldots, k_{b_{m+n}}] \rceil ,$$

where $x_1, \ldots, x_m, x_{m+1}, \ldots, x_{m+n}$ are the first $(m+n)$ variables in alphabetical order.

Proposition 6.3.8 *There is a recursive predicate* $\mathrm{fr}(m, n)$ *such that if m is the Gödel number of a term or a formula E and if n is the Gödel number of a variable v, then*

$$\mathrm{fr}(m, n) \iff v \textit{ is free in } E.$$

Proof. Set

$$\chi^1_{\mathrm{fr}}(m, n) = \chi_{\mathrm{fr}}((m)_1, n) \ldots \chi_{\mathrm{fr}}((m)_{lh(m)-1}, n).$$

Now take

$$\chi_{\mathrm{fr}}(m, n) = \begin{cases} 0 & \text{if } \mathrm{vble}(m) \wedge m = n, \\ \chi^1_{\mathrm{fr}}(m, n) & \text{if } \mathrm{pred}((m)_0) \vee \mathrm{func}((m)_0), \\ \chi_{\mathrm{fr}}((m)_1, n) & \text{if } (m)_0 = SN(\neg), \\ \chi_{\mathrm{fr}}((m)_1, n) \cdot \chi_{\mathrm{fr}}((m)_2, n), & \text{if } lh(m) = 3 \\ & \wedge (m)_0 = SN(\vee), \\ \chi_{\mathrm{fr}}((m)_2, n)), & \text{if } lh(m) = 3 \\ & \wedge (m)_0 = SN(\exists) \wedge (m)_1 \neq n, \\ 1 & \text{otherwise.} \end{cases}$$

Then χ_{fr} is a recursive function with the desired properties. □

Proposition 6.3.9 *There is a recursive function* $\mathrm{substl}(l, m, n)$ *such that if l is the Gödel number of a formula A, if m is the Gödel number of a variable v, and if n is the Gödel number of a term t, then*

$$\mathrm{substl}(l, m, n) \iff t \textit{ is substitutable for } v \textit{ in } A.$$

The proof is left as an exercise.

Exercise 6.3.10 Let I be an interpretation of a theory T' in T. Show that there is a recursive function $f : \mathbb{N} \to \mathbb{N}$ such that if n is the Gödel number of a formula A of T', then $f(n)$ is the Gödel number of its meaning A^I in I.

We shall now show that the set LAx_T of Gödel numbers of all logical axioms of T is recursive. We define four recursive predicates first:

$$\mathrm{pax}(m) \iff \exists^{<m} n[\mathrm{form}(n) \wedge m = \langle SN(\vee), \langle SN(\neg), n\rangle, n\rangle].$$

Note that for all m,

$$\mathrm{pax}(m) \iff m \text{ is the Gödel number of a propositional axiom,}$$

$$\mathrm{idax}(m) \iff \exists^{<m} n[\mathrm{vble}(n) \wedge m = \langle SN(=), n, n\rangle].$$

Note that

$$\mathrm{idax}(m) \iff m \text{ is the Gödel number of an identity axiom.}$$

Exercise 6.3.11 Show that the unary predicates sax and eax of Gödel numbers of all substitution axioms and of all equality axioms respectively are recursive.

We now have the following theorem.

Theorem 6.3.12 *The unary predicate* $\mathrm{LAx}_T \subset \mathbb{N}$ *consisting of Gödel numbers of all logical axioms of* T *is recursive.*

We call a theory T **axiomatized** if the set $\mathrm{NAx}_T \subset \mathbb{N}$ of Gödel numbers of all nonlogical axioms of T is recursive.

That T is axiomatized means that there is an algorithm to decide whether a formula of T is an axiom. This is a natural condition on the set of axioms for any axiomatic system.

We define

$$\mathrm{Ax}_T(n) \iff n \text{ is the Gödel number of an axiom of } T.$$

The following result is obvious from Theorem 6.3.12 and the definition of axiomatized theories.

Proposition 6.3.13 *If* T *is axiomatized,* Ax_T *is recursive.*

Example 6.3.14 The theory N has only finitely many nonlogical axioms. So, N is axiomatized.

Exercise 6.3.15 (i) Show that the set of Gödel numbers of formulas of N of the form

$$A_v[0] \to \forall v(A \to A_v[Sv]) \to A,$$

where A is a formula and v a variable, is recursive.

(ii) Show that Peano arithmetic is axiomatized.
(iii) Show that ZF and ZFC are axiomatized. (Note that the language for set theory has only one binary relation symbol. You only have to show that the set of all Gödel numbers of comprehension axioms is recursive. Show this for replacement axioms also.)

From now on, in this section, T will be a fixed finite theory.

We now introduce some recursive predicates related to rules of inference:

$$\mathrm{cont}(n, m) \Leftrightarrow m = \langle SN(\vee), n, n\rangle.$$

If $n = \lceil B \rceil$ and $m = \lceil B \vee B \rceil$, then $\mathrm{cont}(m, n)$ holds. Further, if n is the Gödel number of a formula B and if m is the Gödel number of a formula A and if $\mathrm{cont}(m, n)$ holds, then A can be inferred from B by the contraction rule.

We define recursive predicates corresponding to other rules as follows:

$$\exp(m, n) \Leftrightarrow \mathrm{form}(n) \wedge m = \langle SN(\vee), (m)_1, n\rangle,$$

$$\begin{aligned}\mathrm{assoc}(m, n) \Leftrightarrow n = &\langle SN(\vee), (n)_1, \langle SN(\vee), ((n)_2)_1, ((n)_2)_2\rangle, \rangle\\ &\wedge (m)_0 = SN(\vee) \wedge ((m)_1)_0 = SN(\vee)\\ &\wedge ((m)_1)_1 = (n)_1\\ &\wedge ((m)_1)_2 = ((n)_2)_1 \wedge (m)_2 = ((n)_2)_2,\end{aligned}$$

$$\begin{aligned}\mathrm{cut}(l, m, n) \Leftrightarrow m = &\langle SN(\vee), (m)_1, (m)_2\rangle\\ &\wedge n = \langle SN(\vee), \langle SN(\neg), (m)_1\rangle, (n)_2\rangle\\ &\wedge l = \langle SN(\vee), (m)_2, (n)_2\rangle,\end{aligned}$$

and

$$\begin{aligned}\mathrm{intr}(m, n) \Leftrightarrow n = &\langle SN(\vee), \langle SN(\neg), ((n)_1)_1\rangle, (n)_2\rangle\\ &\wedge m = \langle SN(\vee), \langle SN(\neg), \langle SN(\exists), (((m)_1)_1)_1, ((n)_1)_1\rangle, (n)_2\rangle\rangle\\ &\wedge \mathrm{vble}(((m)_1)_1) \wedge \neg \mathrm{fr}(((m)_1)_1, (n)_2).\end{aligned}$$

Exercise 6.3.16 Show that the predicates cont, assoc, cut, and intr are recursive.

We continue the idea further. Recall that any proof in a theory is a finite sequence of formulas of the theory. Hence, the following definition is quite important. It will be used to code proofs by numbers in such a way that there is a mechanical procedure to decide whether a natural number codes a proof. Further the procedure decodes the proof.

If $A_1, \ldots, A_n$ is a sequence of formulas of T, the number

$$\langle \lceil A_1 \rceil, \ldots, \lceil A_n \rceil\rangle$$

will be called the Gödel number of the sequence.

Proposition 6.3.17 *If T is axiomatized, the set Pr_T of Gödel numbers of all proofs in T is recursive.*

Proof. This follows from the following equivalence:

$$\begin{aligned}\mathrm{Pr}_T(n) \iff &\mathrm{seq}(n) \wedge \forall i < lh(n)[\mathrm{Ax}_T((n)_i)\\ &\vee \exists j < i((\mathrm{cont} \vee \mathrm{assoc} \vee \mathrm{intr})((n)_i, (n)_j))\\ &\vee \exists j, k < i[\mathrm{cut}((n)_i, (n)_j, (n)_k)].\end{aligned}$$

□

Proposition 6.3.18 *If T is axiomatized, the set $\mathrm{Prf}_T \subset \mathbb{N} \times \mathbb{N}$, of all pairs of numbers (m, n) such that m is the Gödel number of a proof of a formula whose Gödel number is n, is recursive.*

Proof. Note that for any m, n,

$$\mathrm{Prf}_T(m, n) \iff lh(m) > 0 \wedge \mathrm{Pr}_T(m) \wedge (m)_{lh(m) \dot{-} 1} = n.$$

□

Theorem 6.3.19 *If T is axiomatized, then the set Thm_T of Gödel numbers of all theorems of T is semirecursive.*

Proof. This follows from the following equivalence:

$$\mathrm{Thm}_T(n) \iff \exists m[\mathrm{Prf}_T(m, n)].$$

□

But is the predicate Thm_T recursive? Not always. However, in the next section we shall prove that if, moreover, T is complete, then Thm_T is recursive. Quite interestingly, in the next chapter we shall show that Thm_N, Thm_{PA}, and Thm_{ZF} are not recursive. This is the essence of Gödel's first incompleteness theorem.

Remark 6.3.20 Let $F(X_1, \ldots, X_n) \in \mathbb{Z}[X_1, \ldots, X_n]$. We can give a similar coding scheme and assign a natural number, say $g(F)$, to F in such a way that Hilbert's tenth problem has a positive answer if and only if the set H of all those $g(F)$ for which the Diophantine equation $F = 0$ has an integral solution is recursive. We invite readers to carry out such a coding. It has been shown that Hilbert's tenth problem has a negative answer.

6.4 Decidable Theories

We call a finite theory T **decidable** if Thm_T is recursive. Otherwise, the theory is called **undecidable**.

Thus, if a theory is decidable, there is an algorithm to decide whether a formula of T is a theorem or not. Hilbert believed that there should be a decidable set of axioms of number theory (and of most of the interesting mathematical theories) such that every true formula of N is provable. Gödel shocked the mathematical world by showing the impossibility of Hilbert's dream.

Our next result is the following.

Theorem 6.4.1 *Every finite, complete theory is decidable.*

We need the following lemma.

Lemma 6.4.2 *There is a recursive map $g : \mathbb{N} \to \mathbb{N}$ such that if n is the Gödel number of a formula A, then $g(n)$ is the Gödel number of a closed formula B such that*

$$T \vdash A \Longleftrightarrow T \vdash B.$$

Proof. Consider the function $f : \mathbb{N} \times \mathbb{N} \to \mathbb{N}$ defined by

$$f(0, n) = n,$$

and for all $m \geq 0$,

$$f(m + 1, n) = \langle SN(\neg), \langle SN(\exists), \langle 2m\rangle, \langle SN(\neg), f(m, n)\rangle\rangle\rangle.$$

It is routine to check that f is recursive. Further, if n is the Gödel number of a formula A, then $f(m + 1, n)$ is the Gödel number of $\forall x_m \ldots \forall x_0 A$, where $x_0, \ldots, x_m$ are the first $(m + 1)$ variables in alphabetical order. Now set

$$g(n) = f(n, n), \quad n \in \mathbb{N}.$$

Using the closure theorem, generalization rule, and the closure properties of the class of recursive functions and recursive predicates, it is easy to check that the map g has the desired properties. □

Proof of 6.4.1. By Theorems 6.2.4 and 6.3.19, it is sufficient to show that $\neg\mathrm{Thm}_T$ is semirecursive. Fix any $n \in \mathbb{N}$. Now note the following:

$$\begin{aligned} \neg\mathrm{Thm}_T(n) &\Longleftrightarrow \neg\mathrm{form}(n) \vee \mathrm{Thm}_T(\langle SN(\neg), g(n)\rangle) \\ &\Longleftrightarrow \exists m[\neg\mathrm{form}(n) \vee \mathrm{Prf}_T(m, \langle SN(\neg), g(n)\rangle)]. \end{aligned}$$

Since Prf_T is recursive and g is recursive, it follows that $\neg\mathrm{Thm}_T$ is semirecursive. □

This is a very important theorem. It says that if T is axiomatized and Thm_T is not recursive, then T is not complete. Gödel uses it beautifully to establish the first incompleteness theorem.

Exercise 6.4.3 Let $p = 0$ or a prime > 1. Show that the theory $ACF(p)$ is axiomatized and decidable.

This theorem, in particular, says that there is an algorithm to decide whether a statement in the language of ring theory with identity is valid in $\mathbb{C}$.

A simple extension T' of T is called a **finite extension** if at most finitely many nonlogical axioms of T' are not theorems of T.

Theorem 6.4.4 *Let T be an undecidable theory. Suppose T' satisfies one of the following conditions:*

(a) T' is a conservative extension of T.
(b) T is an extension by definition of T'.
(c) T is a finite consistent extension of T'.
(d) T has a faithful interpretation in T'.

Then T' is undecidable.

Proof. In each case, we show that there is a recursive function $f : \mathbb{N} \to \mathbb{N}$ such that if n is the Gödel number of a formula φ of T, $f(n)$ is the Gödel number of a formula φ^* of T' such that

$$T \vdash \varphi \Longleftrightarrow T' \vdash \varphi^*.$$

Assuming this done, we complete the proof first. Suppose in some case (a)–(d), $\mathrm{Thm}_{T'}$ is recursive. Then,

$$\mathrm{Thm}_T(n) \Longleftrightarrow \mathrm{form}_T(n) \wedge \mathrm{Thm}_{T'}(f(n)).$$

Hence, Thm_T is recursive by the closure properties of the set of recursive predicates. This contradicts that T is undecidable.

In case (a), we take $f(n) = n$. Since an extension by definitions of T is a conservative extension of T (Theorem 5.3.6), the result in case (b) follows from case (a).

We now prove the result in case (c). Let $B_1, \ldots, B_m$ be an enumeration of the closures of all the nonlogical axioms of T' which are not theorems of T. Let p be the Gödel number of $\neg B_1 \vee \cdots \vee \neg B_m$. If n is the Gödel number of a formula A of T, we define

$$f(n) = \langle SN(\vee), p, n \rangle.$$

Otherwise, we define $f(n) = 0$. Since the set of Gödel numbers of formulas is a recursive set, it follows that f is recursive. By the reduction theorem (Exercise 4.4.6),

$$\mathrm{Thm}_T(n) \Longleftrightarrow \mathrm{Thm}_{T'}(f(n)).$$

To prove the result in case (d), fix a faithful interpretation I of T in T'. It is fairly routine to see that there is a recursive function $f : \mathbb{N} \to \mathbb{N}$ such that if n is the Gödel number of a formula A of T, then $f(n)$ is the Gödel number of A^I (Exercise 6.3.10). Since I is a faithful interpretation,

$$\mathrm{Thm}_T(n) \iff \mathrm{form}_T(n) \wedge \mathrm{Thm}_{T'}(f(n)).$$

□

7

Incompleteness Theorems and Recursion Theory

This chapter gives the most important landmarks of mathematical logic—the incompleteness theorems of Gödel. We still have to do some work, which we do in the first section. As a side output, in Section 3, we initiate the study of recursion theory.

7.1 Representability

In this section, we present yet another beautiful concept, called *representability*, introduced by Gödel, which shows that recursive functions and predicates can be represented by formulas of the theory N.

Let $P \subset \mathbb{N}^p$. We say that a formula A of N with distinct variables $v_1, \ldots, v_p$ **represents** P if for every sequence of numbers $n_1, \ldots, n_p$,

$$(n_1, \ldots, n_p) \in P \Rightarrow N \vdash A_{v_1, \ldots, v_p}[k_{n_1}, \ldots, k_{n_p}]$$

and

$$(n_1, \ldots, n_p) \notin P \Rightarrow N \vdash \neg A_{v_1, \ldots, v_p}[k_{n_1}, \ldots, k_{n_p}].$$

We say that P is **representable** if some formula A with distinct variables $v_1, \ldots, v_p$ represents it.

Let $f : \mathbb{N}^p \to \mathbb{N}$ be a map. We say that a formula A of N with distinct variables $v_1, \ldots, v_p, w$ **represents** f if for every sequence of numbers $n_1, \ldots, n_p$,

$$N \vdash A_{v_1,\ldots,v_p}[k_{n_1},\ldots,k_{n_p}] \leftrightarrow w = k_m,$$

where $m = f(n_1, \ldots, n_p)$. We say that f is **representable** if some formula A with distinct variables $v_1, \ldots, v_p, w$ represents it.

Theorem 7.1.1 *Every representable predicate $P \subset \mathbb{N}^n$ is recursive.*

Proof. Let a formula φ of N with variables $x_1, \ldots, x_n$ represent P. Then for every $(a_1, \ldots, a_n) \in \mathbb{N}^n$, we have the following:

$$P(a_1,\ldots,a_n) \Rightarrow N \vdash \varphi_{x_1,\ldots,x_n}[k_{a_1},\ldots,k_{a_n}]$$

and

$$\neg P(a_1,\ldots,a_n) \Rightarrow N \vdash \neg\varphi_{x_1,\ldots,x_n}[k_{a_1},\ldots,k_{a_n}].$$

Since N is consistent, we now have the following:

$$P(a_1,\ldots,a_n) \Leftrightarrow \mathrm{Thm}_N(\ulcorner\varphi_{x_1,\ldots,x_n}[k_{a_1},\ldots,k_{a_n}]\urcorner)$$

and

$$\neg P(a_1,\ldots,a_n) \Leftrightarrow \mathrm{Thm}_N(\ulcorner\neg\varphi_{x_1,\ldots,x_n}[k_{a_1},\ldots,k_{a_n}]\urcorner).$$

Since N is axiomatized, by Theorem 6.3.19, Thm_N is semirecursive. Further, the maps

$$(a_1,\ldots,a_n) \to \ulcorner\varphi_{x_1,\ldots,x_n}[k_{a_1},\ldots,k_{a_n}]\urcorner$$

and

$$(a_1,\ldots,a_n) \to \ulcorner\neg\varphi_{x_1,\ldots,x_n}[k_{a_1},\ldots,k_{a_n}]\urcorner$$

are recursive. (See Exercise 6.3.6.) Hence, both P and $\neg P$ are semirecursive. The result now follows from Theorem 6.2.4. □

The main theorem of this section is the following.

Theorem 7.1.2 (Representability theorem) *Every recursive function and every recursive predicate is representable.*

This result is proved essentially by showing that the initial functions are representable and that the set of all representable functions is closed under composition and minimalization. We will see later in this section that every representable function is recursive.

Lemma 7.1.3 *Let P be a p-ary representable predicate on $\mathbb{N}$ and $x_1, \ldots, x_p$ distinct variables. Then there is a formula B of N such that B with $x_1, \ldots, x_p$ represents P.*

Proof. Let A with $v_1, \ldots, v_p$ represent P. Taking a variant of A, if necessary, by the variant theorem, we can assume that $x_1, \ldots, x_p$ do not occur in A. Now take B to be

$$A_{v_1,\ldots,v_p}[x_1, \ldots, x_p].$$

□

The following result is also proved similarly.

Lemma 7.1.4 *Let f be a p-ary representable function on $\mathbb{N}$ and $x_1, \ldots, x_p, y$ distinct variables. Then there is a formula B of N such that B with $x_1, \ldots, x_p, y$ represents f.*

Let f be a p-ary function. We say that a term t of N with distinct variables $v_1, \ldots, v_p$ represents f if for every $n_1, \ldots, n_p$,

$$N \vdash t_{v_1,\ldots,v_p}[k_{n_1}, \ldots, k_{n_p}] = k_m,$$

where $m = f(n_1, \ldots, n_p)$.

Lemma 7.1.5 *Let a term t with distinct variables $v_1, \ldots, v_p$ represent $f : \mathbb{N}^p \to \mathbb{N}$ and let w be a variable distinct from each v_i. Then the formula $w = t$ with $v_1, \ldots, v_p, w$ represents f.*

Proof. Let $m = f(n_1, \ldots, n_p)$. We have

$$N \vdash t_{v_1,\ldots,v_p}[k_{n_1}, \ldots, k_{n_p}] = k_m.$$

We are required to show that

$$N \vdash w = t_{v_1,\ldots,v_p}[k_{n_1}, \ldots, k_{n_p}] \leftrightarrow w = k_m.$$

This essentially follows from the equality axiom, the substitution rule, and the detachment rule. □

Proposition 7.1.6 *A p-ary predicate P is representable if and only if χ_P is representable.*

Proof. Let A with distinct variables $v_1, \ldots, v_p$ represent P. Let the variable w be distinct from each of v_i and let B be the formula

$$(A \wedge w = k_0) \vee (\neg A \wedge w = k_1).$$

Fix any $(n_1, \ldots, n_p)$. Suppose $(n_1, \ldots, n_p) \in P$. Then

$$N \vdash A_{v_1,\ldots,v_p}[k_{n_1}, \ldots, k_{n_p}]. \tag{1}$$

By (1) and the the tautology theorem,

$$N \vdash B_{v_1,\ldots,v_p}[k_{n_1}, \ldots, k_{n_p}] \leftrightarrow w = k_0.$$

If $(n_1, \ldots, n_p) \notin P$,

$$N \vdash \neg A_{v_1,\ldots,v_p}[k_{n_1},\ldots,k_{n_p}]. \tag{2}$$

By (2) and the tautology theorem,

$$N \vdash B_{v_1,\ldots,v_p}[k_{n_1},\ldots,k_{n_p}] \leftrightarrow w = k_1.$$

Conversely, assume that A with distinct variables $v_1,\ldots,v_p,w$ represents χ_P. Let B be the formula $A_w[k_o]$. We claim that B with $v_1,\ldots,v_p$ represents P.

Since $\neg(Sx = 0)$ is an axiom of N,

$$N \vdash \neg(k_1 = k_0).$$

Fix any $(n_1,\ldots,n_p)$. Suppose $(n_1,\ldots,n_p) \in P$. Then

$$N \vdash w = k_0 \leftrightarrow A_{v_1,\ldots,v_p}[k_{n_1},\ldots,k_{n_p}].$$

By the substitution rule,

$$N \vdash k_0 = k_0 \leftrightarrow B_{v_1,\ldots,v_p}[k_{n_1},\ldots,k_{n_p}].$$

Since $N \vdash k_0 = k_0$, we have

$$N \vdash B_{v_1,\ldots,v_p}[k_{n_1},\ldots,k_{n_p}].$$

Since $N \vdash \neg(k_1 = k_0)$, the other case is similarly proved. □

Note that by the previous proposition, to prove the representability theorem, we only need to show that every recursive function is representable. We shall show that all initial functions are representable and that the set of all representable functions is closed under composition and minimalization.

Proposition 7.1.7 *The formula $x = y$ with distinct variables represents $=$ in N.*

Proof. We need to show the following:

$$m = n \Rightarrow N \vdash k_m = k_n$$

and

$$m \neq n \Rightarrow N \vdash \neg(k_m = k_n).$$

The first assertion follows from the identity axiom and the substitution rule. By the symmetry theorem (Lemma 3.5.1), in the proof of the second assertion, we can assume that $m > n$. We proceed by induction on n. If $n = 0$, for all m, this follows from the axiom (1) of N and the substitution rule. Let $m \geq n > 0$ and the second assertion hold for $n - 1$ and all m. Now note the following:

$$N \vdash k_m = k_n \Rightarrow k_{m-1} = k_{n-1},$$

by the axiom (2) of N, the closure theorem, and the substitution rule. By the induction hypothesis, we have

$$N \vdash \neg(k_{m-1} = k_{n-1}).$$

Hence,

$$N \vdash \neg(k_m = k_n),$$

by the tautology theorem. □

Proposition 7.1.8 *All initial functions are representable.*

Proof. (i) Let $v_1, \ldots, v_n$ be distinct variables and let t be the term v_i. Fix natural numbers $p_1, \ldots, p_n$. Clearly

$$N \vdash t_{v_1,\ldots,v_n}[k_{p_1}, \ldots, k_{p_n}] = k_{p_i}.$$

This shows that the projection maps Π_i^n, $n \geq 1$, $1 \leq i \leq n$, are representable.

(ii) Let x and y be distinct variables and let t be the term $x + y$. We show that t with x, y represents $+$. We need to show that for all natural numbers m, n,

$$N \vdash k_m + k_n = k_{m+n}. \tag{1}$$

We fix m and show (1) by induction on n. By the axiom (3) of N, we have

$$N \vdash k_m + 0 = k_m.$$

Now assume that

$$N \vdash k_m + k_n = k_{m+n}. \tag{2}$$

By the axiom (4) of N and the substitution rule, we have

$$N \vdash k_m + k_{n+1} = S(k_m + k_n).$$

By the equality axiom and (2), we have

$$N \vdash S(k_m + k_n) = k_{m+n+1}.$$

So,

$$N \vdash k_m + k_{n+1} = k_{m+n+1}.$$

(iii) Similarly, using axioms (5) and (6) of N, we show that the term $x \cdot y$ with distinct variables x and y represents $\cdot$, the multiplication.

(iv) Finally, we show that the formula $x < y$ with distinct variables x and y represents $<$. This in turn will show that $\chi_<$ is representable and the result will be proved. We are required to show that for every natural numbers m and n,

$$m < n \Longrightarrow N \vdash k_m < k_n \tag{3}$$

and

$$\neg(m < n) \Longrightarrow N \vdash \neg(k_m < k_n). \tag{4}$$

For every n, we show that (3) and (4) hold for all m. We proceed by induction on n. For $n = 0$, we only need to prove (4) for all m; (4) follows from the axiom (7) of N.

Now assume that for some n, (3) and (4) holds for all m. Suppose $m < n + 1$. If $m < n$, then

$$N \vdash k_m < k_n$$

and hence

$$N \vdash k_m < k_{n+1}$$

by the axiom (8), the substitution rule, and the induction hypothesis. If $m = n$,

$$N \vdash k_m = k_n$$

by Proposition 7.1.7. Then

$$N \vdash k_m < k_{n+1}$$

by the same arguments.

Now suppose $m \geq n + 1$. Then,

$$N \vdash \neg(k_m < k_n)$$

by the induction hypothesis and

$$N \vdash \neg(k_m = k_n)$$

by Proposition 7.1.7. So,

$$N \vdash \neg(k_m < k_{n+1})$$

using the axiom (8) of N and the tautology theorem. □

Proposition 7.1.9 *The set of all representable functions is closed under composition.*

Proof. Let

$$h(n_1, \ldots, n_k) = g(f_1(n_1, \ldots, n_k), \ldots, f_m(n_1, \ldots, n_k)),$$

where g, $f_1, \ldots, f_m$ are representable. We choose distinct variables, $u, v_1, \ldots, v_m, w_1, \ldots, w_k$ and formulas B, $A_1, \ldots, A_m$ such that B with $v_1, \ldots, v_m$ and u represents g and A_i with $w_1, \ldots, w_k$ and v_i represents f_i, $1 \leq i \leq m$.

Now consider the formula C defined by

$$\exists v_1 \cdots \exists v_m (A_1 \wedge \cdots \wedge A_m \wedge B).$$

We claim that C with $w_1, \ldots, w_k$ and u represents h. Fix $(n_1, \ldots, n_k)$. Let $p_i = f_i(n_1, \ldots, n_k)$ and $q = g(p_1, \ldots, p_m)$. Then $q = h(n_1, \ldots, n_k)$. For $1 \leq i \leq m$, set

$$A'_i = (A_i)_{w_1,\ldots,w_k}[k_{n_1}, \ldots, k_{n_k}]$$

and

$$C' = C_{w_1,\ldots,w_k}[k_{n_1}, \ldots, k_{n_k}].$$

By our assumptions, we have

$$N \vdash A'_i \leftrightarrow v_i = k_{p_i},$$

$1 \leq i \leq m$. By the equivalence theorem, we have

$$N \vdash C' \leftrightarrow \exists v_1 \cdots \exists v_m (v_1 = k_{p_1} \wedge \cdots \wedge v_m = k_{p_m} \wedge B).$$

Let D denote the formula

$$\exists v_1 \cdots \exists v_m (v_1 = k_{p_1} \wedge \cdots \wedge v_m = k_{p_m} \wedge B).$$

By the repeated application of Proposition 4.2.23, we have

$$N \vdash \exists v_2 \cdots \exists v_m (v_2 = k_{p_2} \wedge \cdots \wedge v_m = k_{p_m} \wedge B_{v_1}[k_{p_1}]) \leftrightarrow D,$$

$$\vdots$$

$$N \vdash B_{v_1,\ldots,v_m}[k_{p_1}, \ldots, k_{p_m}] \leftrightarrow \exists v_m (v_m = k_{p_m} \wedge B_{v_1,\ldots,v_{m-1}}[k_{p_1}, \ldots, k_{p_{m-1}}]).$$

By the equivalence theorem and the tautology theorem, we get

$$N \vdash C' \leftrightarrow B_{v_1,\ldots,v_m}[k_{p_1}, \ldots, k_{p_m}].$$

Since B with $v_1, \ldots, v_m$ and u represents g and $q = g(p_1, \ldots, p_m)$, we have

$$N \vdash B_{v_1,\ldots,v_m}[k_{p_1}, \ldots, k_{p_m}] \leftrightarrow u = k_q.$$

Thus by the equivalence theorem,

$$N \vdash C' \leftrightarrow u = k_q.$$

□

It remains to show that the set of all representable functions is closed under minimalization.

Proposition 7.1.10 *The set of all representable functions is closed under minimalization.*

Proof. Let $f(m,\overline{n})$ be representable, where $\overline{n} = (n_0, \ldots, n_{p-1})$. Let A with $v, v_0, \ldots, v_{p-1}$ and w represent f. Assume that

$$\forall \overline{n} \exists m (f(m,\overline{n}) = 0).$$

Let

$$g(\overline{n}) = \mu m (f(m,\overline{n}) = 0).$$

We now show that g is representable.

Let u be a new variable and let B be the formula

$$A_w[0] \wedge \forall u (u < v \to \neg A_{v,w}[u,0]).$$

We claim that B with $v_0, \ldots, v_{p-1}$ and v represents g.

Fix $\overline{n} = (n_0, \ldots, n_{p-1}) \in \mathbb{N}^p$. Let $m = g(\overline{n})$. Then $f(i,\overline{n}) = l_i \neq 0$, $i < m$, and $f(m,\overline{n}) = 0$. Set

$$A' = A_{v_0,\ldots,v_{p-1}}[k_{n_0}, \ldots, k_{n_{p-1}}]$$

and

$$B' = B_{v_0,\ldots,v_{p-1}}[k_{n_0}, \ldots, k_{n_{p-1}}].$$

We have

$$N \vdash A'_v[k_i] \leftrightarrow w = k_{l_i},$$

$i < m$, and

$$N \vdash A'_v[k_m] \leftrightarrow w = 0.$$

Since $l_i \neq 0$, we have

$$N \vdash \neg(0 = k_{l_i}),$$

$i < m$. Thus, for all $i < m$, we have

$$N \vdash \neg A'_{v,w}[k_i, 0]$$

and

$$N \vdash A'_{v,w}[k_m, 0].$$

Hence, by Proposition 4.3.2,

$$N \vdash (A'_w[0] \wedge \forall u (u < v \to \neg A'_{v,w}[u,0])) \to v = k_m,$$

i.e.,

$$N \vdash B' \leftrightarrow v = k_m.$$

Our claim is proved. □

We have completed the proof of the representability theorem.

Exercise 7.1.11 Show that every representable function $f : \mathbb{N}^n \to \mathbb{N}$ is recursive.

Remark 7.1.12 We have presented an amazing loop constructed by Gödel that has been likened to the music of Bach and the drawings of Escher [4]. To make statements about numbers, first one develops a formal language (for instance, the language of N) and expresses statements about numbers syntactically in N (or in a suitable extension of N). In this language, a statement about numbers is now a sentence of N. Then using the idea of Gödel numbers, one expresses statements about syntactical objects by numbers themselves. Finally, by the representability theorem, one represents a certain class S of statements about numbers by formulas of N. For instance, if

$$\{n \in \mathbb{N} : n \text{ is the Gödel number of a sentence in } S\}$$

is recursive, we represent it by a formula of N. This technique enables one to hop into the theory N from the metaworld and vice versa. Thus many questions in the metaworld are expressed by formulas of N. Sometimes even a proof in the metaworld is converted into a proof inside the theory. Thus, Gödel built a very powerful tool and destroyed the beliefs of many great mathematicians of his time, including Hilbert. In the remaining part of this chapter, we present such remarkable discoveries of Gödel.

7.2 First Incompleteness Theorem

Theorem 7.2.1 (First incompleteness theorem) *Every axiomatized, consistent extension of N is undecidable and so incomplete.*

Proof. Toward arriving at a contradiction, assume that Thm_T is recursive. Fix a variable v. There is a recursive function $f : \mathbb{N} \times \mathbb{N} \to \mathbb{N}$ such that if m is the Gödel number of a formula B of T, then $f(m, n)$ is the Gödel number of $B_v[k_n]$. Then the binary predicate $U \subset \mathbb{N} \times \mathbb{N}$ defined by

$$U(m, n) \Leftrightarrow \mathrm{Thm}_T(f(m, n))$$

is recursive. So, the predicate

$$P(m) \Leftrightarrow \neg U(m, m)$$

is recursive.

By the representability theorem, there is a formula A of N such that A with v represents P. This means that for every $n \in \mathbb{N}$,

$$n \in P \Rightarrow N \vdash A_v[k_n]$$

and

$$n \notin P \Rightarrow N \vdash \neg A_v[k_n].$$

Since T is an extension of N, A is a formula of T and for every $n \in \mathbb{N}$,

$$n \in P \Rightarrow T \vdash A_v[k_n] \tag{a}$$

and

$$n \notin P \Rightarrow T \vdash \neg A_v[k_n]. \tag{b}$$

Now let m be the Gödel number of A.

Suppose $m \in P$. Then, by (a), $T \vdash A_v[k_m]$, i.e., $U(m, m)$ holds by the definition of U. Hence, $m \notin P$ by the definition of P. This is a contradiction.

On the other hand, suppose $m \notin P$. Then, by (b), $T \vdash \neg A_v[k_m]$. Since T is consistent, this implies that $T \nvdash A_v[k_m]$. Therefore, by the definition of U, $\neg U(m, m)$ holds. But then $m \in P$ by the definition of P. We have arrived at a contradiction again. □

Remark 7.2.2 Since $\mathbb{N}$ with the usual interpretations of S, $+$, $\cdot$, and $<$ is a model of PA, PA is consistent. Further, by Exercise 6.3.15, PA is axiomatized. Hence, by Theorem 6.4.4, PA is undecidable, and so incomplete by the first incompleteness theorem.

Remark 7.2.3 There is an extension by definitions of ZF (or of ZFC) in which there is a suitable interpretation of the theory N so that we can carry out the same arguments and prove the incompleteness of ZF and ZFC.

Remark 7.2.4 Comparison with the liar's paradox. In the above proof we have produced a formula $A[v]$ that says that the formula whose Gödel number is m is not provable for "$v = m$." This is similar to the statement "I am lying" of the liar's paradox. This argument also shows a way to make a self-referential statement inside a theory.

Remark 7.2.5 Now we can state Hilbert's problem precisely: is there an axiomatized extension P' of PA such that a sentence of P' is true (in the standard model) if and only if it is a theorem of P'. Since such a theory P' is complete, the first incompleteness theorem answers Hilbert's question in the negative.

7.3 Arithmetical Sets

In this section, we present some basic results in recursion theory. We have already initiated the study of recursive and semirecursive sets. The representability theorem and the ideas contained in the proof of the first incompleteness theorem help us to continue this study further.

A set of predicates (of not necessarily fixed arity) on $\mathbb{N}$ will be called a **pointclass**. For brevity, we shall write $\overline{n}$ for $(n_1, \ldots, n_k)$. For $P \subset \mathbb{N} \times \mathbb{N}^k$, we define k-ary predicates $\exists^\omega P$ and $\forall^\omega P$ by

$$\exists^\omega P(\overline{n}) \iff \exists m P(m, \overline{n})$$

and

$$\forall^\omega P(\overline{n}) \iff \forall m P(m, \overline{n}).$$

If Γ is a pointclass, we define

$$\neg\Gamma = \{\neg P : P \in \Gamma\},$$

$$\exists^\omega \Gamma = \{\exists^\omega P : P \in \Gamma\},$$

and

$$\forall^\omega \Gamma = \{\forall^\omega P : P \in \Gamma\}.$$

Note that for any pointclass,

$$\forall^\omega \Gamma = \neg\exists^\omega \neg\Gamma$$

and

$$\exists^\omega \Gamma = \neg\forall^\omega \neg\Gamma.$$

In the sequel, these two identities and other such simple set-theoretic identities will be used without mention.

We define **arithmetical pointclasses** Σ_n^0, Π_n^0, and Δ_n^0, $n \geq 1$, by induction as follows:

$$\Sigma_1^0 = \text{ the poinclass of all semirecursive sets,}$$

$$\Pi_n^0 = \neg\Sigma_n^0,$$

$$\Sigma_{n+1}^0 = \exists^\omega \Pi_n^0,$$

and

$$\Delta_n^0 = \Sigma_n^0 \cap \Pi_n^0.$$

Theorem 7.3.1 *The pointclass Δ_1^0 consists precisely of all recursive sets and*

$$\Sigma_1^0 = \exists^\omega \Delta_1^0 \text{ and } \Pi_1^0 = \forall^\omega \Delta_1^0.$$

Proof. The first assertion is just a restatement of Kleene's theorem (Theorem 6.2.4); the second one follows from the definition of semirecursive sets and Kleene's theorem (Theorem 6.2.4), and the third one follows from the second one. □

Call a pointclass Γ closed under **recursive substitutions** if whenever $P(n_1, \ldots, n_k) \in \Gamma$, for all recursive functions $f_i : \mathbb{N}^l \to \mathbb{N}$, $1 \leq i \leq k$, the l-ary predicate Q defined by

$$Q(m_1, \ldots, m_l) \iff P(f_1(m_1, \ldots, m_l), \ldots, f_k(m_1, \ldots, m_l))$$

is in Γ.

Proposition 7.3.2(*1*) *Each arithmetical pointclass is closed under* $\vee$, $\wedge$ *(so, under finite unions and finite intersections), and recursive substitutions.*

(*2*) *For each* n, Σ_n^0 *is closed under* $\exists^\omega$, Π_n^0 *under* $\forall^\omega$, *and* Δ_n^0 *under* $\neg$.

(*3*) *For each* n,

$$\Sigma_n^0 = \exists^\omega \Delta_n^0 \text{ and } \Pi_n^0 = \forall^\omega \Delta_n^0,$$

$$\Sigma_n^0 \cup \Pi_n^0 \subset \Delta_{n+1}^0.$$

Proof. We shall prove (1) by induction on n. The closure properties listed in (1) have already been proved for Σ_1^0 in Proposition 6.2.2. Then the result for Π_1^0 follows from its definition. The result for Δ_1^0 is now easily seen.

Assume that Σ_n^0, Π_n^0, and Δ_n^0 have the closure properties listed in (1). Note that a pointclass Γ is closed under $\exists^\omega$ if and only if $\neg\Gamma$ is closed under $\forall^\omega$; a pointclass Γ is closed under recursive substitutions if and only if $\neg\Gamma$ is and if both Γ and $\neg\Gamma$ satisfy the closure properties listed in (1), so does $\Delta = \Gamma \cap \neg\Gamma$.

Now let $f_i : \mathbb{N}^l \to \mathbb{N}$, $1 \le i \le k$, be recursive functions. Let $P(m_1, \ldots, m_k) \in \Pi_{n+1}^0$. We are required to show that the predicate Q defined by

$$Q(\bar{l}) \iff P(f_1(\bar{l}), \ldots, f_k(\bar{l}))$$

is in Π_{n+1}^0. Get $P' \in \Sigma_n^0$ such that $P = \forall^\omega P'$. Then,

$$Q(\bar{l}) \iff \forall m P'(m, f_1(\bar{l}), \ldots, f_k(\bar{l})).$$

Since Σ_n^0 is closed under recursive substitutions, $Q \in \Pi_{n+1}^0$.

Let $P, Q \in \Sigma_{n+1}^0$. Get $P', Q' \in \Pi_n^0$ such that $P = \exists^\omega P'$ and $Q = \exists^\omega Q'$. The following identities are easy to check:

$$P \vee Q \iff \exists^\omega (P' \vee Q')$$

and

$$(P \wedge Q)(\bar{l}) \iff \exists m (P'((m)_0, \bar{l}) \wedge Q'((m)_1, \bar{l}).$$

By the induction hypothesis, it follows that Σ_{n+1}^0 is closed under $\vee$ and $\wedge$. This, in turn, implies that Π_{n+1}^0 and Δ_{n+1}^0 are closed under $\vee$ and $\wedge$.

We shall now prove (2). The pointclass Δ_n^0 is clearly closed under $\neg$. By Proposition 6.2.2, Σ_1^0 is closed under $\exists^\omega$. Let $P(p, q, \bar{l}) \in \Pi_n^0$. Then

$$\exists p \exists q P(p, q, \bar{l}) \iff \exists m P((m)_0, (m)_1, \bar{l})).$$

Since Π_n^0 is closed under recursive substitutions, it follows that Σ_{n+1}^0 is closed under $\exists^\omega$. Hence, Π_{n+1}^0 is closed under $\forall^\omega$.

We prove (3) also by induction on n. By Propositions 6.2.1 and 6.2.2,

$$\Sigma_1^0 = \exists^\omega \Delta_1^0.$$

So,

$$\Pi_1^0 = \forall^\omega \Delta_1^0.$$

Further, Since $\Delta_1^0 \subset \Pi_1^0$, we conclude that $\exists^\omega \Delta_1^0 \subset \exists^\omega \Pi_1^0$. Hence, $\Sigma_1^0 \subset \Sigma_2^0$ by the definition of Σ_2^0.

If P is Σ_1^0 and

$$Q(m, \overline{n}) \iff P(\overline{n}),$$

then by recursive substitutions, Q is Σ_1^0 and

$$P(\overline{n}) \iff \forall m Q(m, \overline{n}),$$

showing that P is Π_2^0. Thus, $\Sigma_1^0 \subset \Pi_2^0$. This shows that $\Sigma_1^0 \subset \Delta_2^0$. Since Δ_2^0 is closed under $\neg$, it follows that $\Pi_1^0 \subset \Delta_2^0$ as well. This proves the result for $n = 1$. (3) can be proved similarly for all n by induction. □

Remark 7.3.3 The inclusion relations between these pointclasses are summed up by the following diagram:

$$\begin{array}{ccccccc} & \Sigma_1^0 & & \Pi_2^0 & & \Sigma_3^0 & \ldots \\ \Delta_1^0 & & \Delta_2^0 & & \Delta_3^0 & & \ldots \\ & \Pi_1^0 & & \Sigma_2^0 & & \Pi_3^0 & \ldots \end{array}$$

where a pointclass in any column is contained in all pointclasses to its right.

Corollary 7.3.4 *(i) Let n be even. Then*

$$\Sigma_n^0 = \underbrace{\exists^\omega \forall^\omega \cdots \exists^\omega \forall^\omega}_{n \text{ times}} \Delta_1^0$$

and

$$\Pi_n^0 = \underbrace{\forall^\omega \exists^\omega \cdots \forall^\omega \exists^\omega}_{n \text{ times}} \Delta_1^0.$$

(ii) Let n be odd. Then

$$\Sigma_n^0 = \underbrace{\exists^\omega \forall^\omega \cdots \exists^\omega}_{n \text{ times}} \Delta_1^0$$

and

$$\Pi_n^0 = \underbrace{\forall^\omega \exists^\omega \cdots \forall^\omega}_{n \text{ times}} \Delta_1^0.$$

A predicate in

$$\cup_n \Sigma_n^0 = \cup_n \Delta_n^0 = \cup_n \Pi_n^0$$

is called an **arithmetical set**.

Exercise 7.3.5 Show that all the arithmetical pointclasses are closed under $\exists^<$, $\exists^\leq$, $\forall^<$, and $\forall^\leq$.

Using the representability theorem and ideas contained in the proof of the first incompleteness theorem, we now show that all the inclusions in the arithmetical hierarchy are strict.

Let Γ be an arithmetical pointclass. Let $k \geq 1$. Call a $(k+1)$-ary predicate U^k **universal** for Γ if $U^k \in \Gamma$ and if for every k-ary predicate P in Γ there is an m such that for all $\overline{n} \in \mathbb{N}^k$,

$$P(\overline{n}) \Longleftrightarrow U^k(m, \overline{n}).$$

Strictly speaking, we should say that U^k is universal for k-ary predicates in Γ. We shall not do it. It should be understood from the exponent k.

We make some simple observations now.

(a) U^k is universal for Σ^0_n if and only if $\neg U^k$ is universal for Π^0_n.

(b) Let U^{k+1} be universal for Π^0_n. Set

$$U^k = \exists^\omega U^{k+1}.$$

We claim that U^k is universal for Σ^0_{n+1}.

Clearly $U^k \in \Sigma^0_{n+1}$. Now let P be a k-ary predicate in Σ^0_{n+1}. Then there is a $(k+1)$-ary predicate Q in Π^0_n such that

$$P = \exists^\omega Q.$$

Since U^{k+1} is universal for Π^0_n, there is a p such that for all q and for all $\overline{m}$,

$$Q(q, \overline{m}) \Longleftrightarrow U^{k+1}(p, q, \overline{m}).$$

Our assertion follows.

(c) Let U^1 be universal for Σ^0_n. Define P by

$$P(m) \Longleftrightarrow U^1(m, m).$$

Then P is Σ^0_n. We claim that it is not Δ^0_n.

Since Σ^0_n is closed under recursive substitutions, $P \in \Sigma^0_n$. We claim that $P \notin \Pi^0_n$. Suppose not. Then $\neg P \in \Sigma^0_n$. Let m be such that for all k,

$$\neg P(k) \Longleftrightarrow U^1(m, k).$$

But then

$$P(m) \Longleftrightarrow U^1(m, m) \Longleftrightarrow \neg P(m),$$

and we have arrived at a contradiction.

(d) Arguing as in (c), we see that for any n, Δ^0_n does not contain any universal set for Σ^0_n or Π^0_n.

Theorem 7.3.6 *For each $k \geq 1$, there is a $(k+1)$-ary predicate U^k that is universal for Σ_1^0.*

Proof. Let $x_0, x_1, x_2, x_3, \dots$ be all the variables of the theory N in alphabetical order. For each $\overline{a} \in \mathbb{N}^k$, set

$$\mathrm{num}(\overline{a}) = (\mathrm{num}(a_0), \dots, \mathrm{num}(a_{k-1})).$$

Define U^k by

$$U^k(m, \overline{a}) \Longleftrightarrow \exists p \mathrm{Thm}_N(\ulcorner m, \mathrm{num}(p), \mathrm{num}(\overline{a}))),$$

where sb is the recursive function defined in Exercise 6.3.6.

Since $\mathrm{Thm}_N \in \Sigma_1^0$, $U^k \in \Sigma_1^0$. Now let $P \subset \mathbb{N}^k$ be semirecursive. Then there is a recursive set $Q \subset \mathbb{N}^{k+1}$ such that

$$P \Longleftrightarrow \exists^\omega Q.$$

By the representability theorem, there is formula A of N such that A with $x_0, \dots, x_k$ represents Q. Let $m = \ulcorner A \urcorner$. Since N is consistent, we see that for all $\overline{a}$,

$$P(\overline{a}) \Longleftrightarrow U^k(m, \overline{a}).$$

□

The following theorem is also now easy to see.

Theorem 7.3.7 *Let Γ be any of the pointclasses Σ_n^0 or of Π_n^0, $n \geq 1$. Then, for every $k \geq 1$, there is a $(k+1)$-ary predicate $U^k \in \Gamma$ that is universal for Γ.*

Remark 7.3.8 We now see that the hierarchy of arithmetical sets is strict, i.e., for all n, Δ_n^0 is properly contained in both Σ_n^0 and Π_n^0.

Hint: Suppose every Δ_n^0 set is Σ_n^0. Let U^1 be a universal Σ_n^0 set. Then it is a universal Δ_n^0 set too. But such a set cannot exist. (See the proof of Proposition 6.1.28.)

Proposition 7.3.9 *Let Γ be any of Σ_n^0 or of Π_n^0, $n \geq 1$. Then, for every $k \geq 1$, there are $(k+1)$-ary predicates $U_0^k, U_1^k \in \Gamma$ such that for every pair of sets $A_0, A_1 \subset \mathbb{N}^k$ in Γ, there is an m such that*

$$(\forall \overline{a} \in \mathbb{N}^k)(A_0(\overline{a}) \Leftrightarrow U_0^k(m, \overline{a}) \ \wedge \ A_1(\overline{a}) \Leftrightarrow U_1^k(m, \overline{a})).$$

Proof. Define

$$U_i^k(m, \overline{a}) \Longleftrightarrow U^k((m)_i, \overline{a}),$$

$i = 0, 1$. Since Γ is closed under recursive substitutions, each U_0^k and U_1^k is in Γ. Given $A_0, A_1 \subset \mathbb{N}^k$ in Γ, choose m_0, m_1 such that

$$\forall \overline{a}(A_i(\overline{a}) \Longleftrightarrow U^p(m_i, \overline{a})),$$

$i = 0, 1$. Take $m = \langle m_0, m_1 \rangle$. $\square$

The pair U_0^k, U_1^k obtained above will be called a **universal pair for Γ.**

Let Γ be any of Σ_n^0 or of Π_n^0, $n \geq 1$. We say that Γ has the **uniformization propery** if for every $k \geq 1$ and every $P \in \Gamma$, $P \subset \mathbb{N}^k \times \mathbb{N}$, there is a $Q \subset P$ in Γ such that

$$(\forall \overline{a} \in \mathbb{N}^k)(\exists m P(\overline{a}, m) \Rightarrow \exists ! m Q(\overline{a}, m)),$$

where $\exists ! m \ldots$ abbreviates "there is a unique $m \ldots$." Such a set Q is called a **uniformization** of P.

Let Γ be any of Σ_n^0 or of Π_n^0, $n \geq 1$. We say that Γ has the **reduction property** if for every $P_1, P_2 \subset \mathbb{N}^k$ in Γ there exist $Q_1 \subset P_1, Q_2 \subset P_2$ in Γ such that $Q_1 \cap Q_2 = \emptyset$ and $Q_1 \cup Q_2 = P_1 \cup P_2$.

Proposition 7.3.10 *If Γ has the uniformization property, then it has the reduction property.*

Proof. To see the result, given $P_1, P_2 \subset \mathbb{N}^k$ in Γ, define

$$P = (P_1 \times \{1\}) \cup (P_2 \times \{2\}).$$

Then $P \in \Gamma$. Choose a uniformization $Q \subset P$ of P in Γ. Set

$$Q_i(\overline{a}) \iff Q(\overline{a}, i),$$

$i = 1, 2$. $\square$

Let Γ be any of Σ_n^0 or of Π_n^0, $n \geq 1$ and $\Delta = \Gamma \cap \neg\Gamma$. We say that Γ has the **separation property** if for every disjoint $P_1, P_2 \subset \mathbb{N}^k$ in Γ there exists a $Q \subset \mathbb{N}^k$ in Δ such that

$$P_1 \subset Q \ \wedge \ Q \cap P_2 = \emptyset.$$

Proposition 7.3.11 *If Γ has the reduction property, then $\neg\Gamma$ has the separation property.*

Proof. To see this, take $P_1, P_2 \subset \mathbb{N}^k$ in $\neg\Gamma$ such that $P_1 \cap P_2 = \emptyset$. Let

$$Q_i = \mathbb{N}^k \setminus P_i,$$

$i = 1, 2$. Then $Q_1, Q_2 \in \Gamma$ and $Q_1 \cup Q_2 = \mathbb{N}^k$. Since Γ has the reduction property, there exist $R_i \subset Q_i$ in Γ, $i = 1, 2$, such that $R_1 \cap R_2 = \emptyset$ and $R_1 \cup R_2 = \mathbb{N}^k$. Thus $R_1 \in \Delta$. Set $Q = \mathbb{N}^k \setminus R_1$. $\square$

Exercise 7.3.12 Let $\Gamma = \Sigma_n^0$ or Π_n^0, $n \geq 1$. Show that Γ cannot satisfy both the reduction property and the separation property.

Hint: Using Proposition 7.3.9, take a universal pair U_1^1, U_2^1 for Γ. Let $V_1, V_2 \in \Gamma$ reduce the pair U_1^1, U_2^1 in the above sense. Let $W \supset V_1$, $W \cap V_2 = \emptyset$, and $W \in \Delta$. Show that W is universal for Δ.

Theorem 7.3.13 (Uniformization theorem.) *Every arithmetical pointclass Σ_n^0 has the uniformization property.*

Proof. Let $P \subset \mathbb{N}^k \times \mathbb{N}$ be in Σ_n^0. Choose $R \subset \mathbb{N} \times \mathbb{N}^k \times \mathbb{N}$ in Δ_n^0 such that $P = \exists^\omega R$. Define Q by

$$Q'(\overline{a}, n) \iff R((n)_0, \overline{a}, (n)_1) \wedge \forall^{<n} k \neg R((k)_0, \overline{a}, (k)_1)$$

and set

$$Q(\overline{a}, m) \iff \exists n[m = (n)_0 \wedge Q'(\overline{a}, n)].$$

□

Corollary 7.3.14 *Each arithmetical pointclass Π_n^0, $n \geq 1$, has the separation property and each Σ_n^0, $n \geq 1$, has the reduction property*

We close this section by proving two basic results in recursion theory. We shall set up some notation first. For $\overline{a} = (a_1, \ldots, a_m, a_{m+1}, \ldots, a_{m+n})$, we have

$$l_m(\overline{a}) = (a_1, \ldots, a_m)$$

and

$$r_n(\overline{a}) = (a_{m+1}, \ldots, a_{m+n})$$

Theorem 7.3.15 (s_n^m-Theorem) *The sequence of universal sets $U^1, U^2, \ldots$ for Σ_1^0 defined in 7.3.6 satisfy the following:*

$$U^{m+n}(p, \overline{a}) \iff U^m(s_n^m(p, r_n(\overline{a})), a_1, \ldots, a_m),$$

where $\overline{a} = (a_1, \ldots, a_m, a_{m+1}, \ldots, a_{m+n})$ and the function s_n^m is as defined in Exercise 6.3.7.

Proof. The result follows directly from the definitions of the U^k's and the function s_n^m. □

If P is a k-ary semirecursive predicate and m is such that for all $\overline{a} \in \mathbb{N}^k$,

$$P(\overline{a}) \iff U^k(m, \overline{a}),$$

we say that m is a **code** of P.

As an application of the s_n^m-theorem we show the following:

Proposition 7.3.16 *There exist recursive functions $\vee^k(m, n)$ and $\wedge^k(m, n)$ such that if m is a code of $P \subset \mathbb{N}^k$ and n a code of $Q \subset \mathbb{N}^k$, then $\vee^k(m, n)$ and $\wedge^k(m, n)$ are codes of $P \vee Q$ and $P \wedge Q$ respectively.*

Proof. First we define $\vee^k(m,n)$. Let U^k's be as defined in Theorem 7.3.6. Now define

$$R(\overline{a},m,n) \iff U^k(m,\overline{a}) \vee U^k(n,\overline{a}).$$

Then there is a p such that

$$R(\overline{a},m,n) \iff U^{k+2}(p,\overline{a},m,n).$$

Set

$$\vee^k(m,n) = s_2^k(p,m,n).$$

We define $\wedge^k$ similarly. □

We refer to the above closure properties of Σ_1^0 by saying that Σ_1^0 is **uniformly closed** under $\vee$ and $\wedge$ (with respect to the universal sets U^k).

Exercise 7.3.17 Show that Σ_1^0 and Π_1^0 are uniformly closed under $\exists^<$, $\forall^<$, $\exists^\leq$, and $\forall^\leq$. Further, Σ_1^0 is uniformly closed under $\exists^\omega$ and Π_1^0 under $\forall^\omega$. For instance, show that for each $k \geq 1$, there is a recursive function $\exists_k^< : \mathbb{N} \to \mathbb{N}$ such that if n codes $P(m,\overline{a})$, a $(k+1)$-ary predicate, $\exists_k^<(n)$ codes the predicate

$$Q(p,\overline{a}) \iff \exists^{<p} P(p,\overline{a}).$$

Theorem 7.3.18 (Kleene's recursion theorem) *Let P be a $(k+1)$-ary predicate in Σ_1^0. Then there is a n^* such that for all $\overline{m}$,*

$$P(n^*,\overline{m}) \iff U^k(n^*,\overline{m}).$$

Proof. Define a $(k+1)$-ary predicate Q by

$$Q(\overline{m},p) \Leftrightarrow P(s_1^k(p,p),\overline{m}).$$

Then $Q \in \Sigma_1^0$. So, there there is a q such that

$$Q(\overline{m},p) \iff U^{k+1}(q,\overline{m},p).$$

Take $n^* = s_1^k(q,q)$. □

Kleene's recursion theorem is very useful in showing that certain functions and predicates are recursive.

Example 7.3.19 Let α be a unary recursive function and β and γ 3-ary recursive functions. Then the 2-ary function δ defined by

$$\begin{aligned} \delta(0,n) &= \alpha(n), \\ \delta(m+1,n) &= \beta(\delta(m,\gamma(\delta(m,n),m,n)),m,n)), \end{aligned}$$

is recursive.

To show this, by Proposition 6.1.27, it suffices to show that the graph of δ is semirecursive. To show this, we define a 4-ary semirecursive predicate G as follows:

$$\begin{aligned} G(l,m,n,k) \iff & (m = 0 \wedge k = \alpha(n)) \\ & \vee \exists p \exists q \exists r \exists s (m = p + 1 \\ & \wedge U^3(l,p,n,q) \\ & \wedge r = \gamma(q,p,n) \\ & \wedge U^3(l,p,r,s) \\ & \wedge k = \beta(s,p,n)), \end{aligned}$$

where U^3 is the universal set for Σ^0_1-sets in $\mathbb{N}^3$ defined earlier. Clearly, G is a 4-ary semirecursive predicate. Hence, by Kleene's recursion theorem,

$$\exists l^* \forall m \forall n \forall k (G(l^*,m,n,k) \iff U^3(l^*,m,n,k)).$$

It is fairly routine to check that the 3-ary semirecursive predicate $U^3(l^*,\cdot,\cdot,\cdot)$ is the graph of δ.

Exercise 7.3.20 Let $n \geq 1$, and let Γ equal Σ^0_n or Π^0_n, and let the U^k's be the universal sets for Γ obtained in Theorem 7.3.7. Show that the s^m_n-theorem (with the same function s^m_n) and Kleene's recursion theorem hold for Γ.

Remark 7.3.21 It is fairly easy to see that the s^m_n-theorem for any of these arithmetical pointclasses can be used to show their uniform closure properties, and Kleene's recursion theorem can be used to show that predicates are in these classes.

7.4 Recursive Extensions of Peano Arithemetic

The arithmetization of theories due to Gödel enables one to examine questions about a theory, such as PA or ZF, within the theory itself. For instance, using the representability theorem, now we can express the metasentence "Peano arithmetic is consistent" by a formula of PA itself and can examine whether this formula is a theorem of PA. This involves formalizing proofs in metatheory inside the theory itself. A key step toward this is to show that every true closed existential formula of PA is a theorem of PA. (Recall that a sentence φ of the language for N is called true if it is valid in the standard model $\mathbb{N}$ of the theory N.) In this section we prove this vital theorem.

Let P' be an extension by definitions of PA and let φ be a formula of P' in which no variable other than $v_1, \ldots, v_n$ and w are free, $v_1, \ldots, v_n$, w distinct. Suppose $P' \vdash \exists w \varphi$. Let w' be a new variable and ψ the formula

$$\varphi \wedge \forall w'(w' < w \rightarrow \neg \varphi_w[w']).$$

By Exercise 4.3.5, we have the following:

(a) $P' \vdash \exists w \psi$.
(b) $P' \vdash \psi \wedge \psi_w[w''] \to w = w''$.

Thus, we can introduce to P' a new n-ary function symbol f with the defining axiom ψ. We shall write

$$fv_1 \cdots v_n = \mu w \varphi$$

to express that f has been introduced as above with ψ as its defining axiom.

We say that P' is a **recursive extension of** PA of PA if it is obtained by a finite number of extensions of PA where the defining axiom for a predicate is an open formula and the defining axiom for a function symbol is a formula of the form ψ described above with φ open.

Example 7.4.1 Let $1 \leq i \leq n$ and let $\varphi[w, v_1, \ldots, v_n]$ be the formula $w = v_i$. Then

$$\pi_i^n v_1 \cdots v_n = \mu w \varphi$$

introduces the projection map π_i^n in a recursive extension of PA.

The following exercise is quite easy to prove.

Exercise 7.4.2 Show that functions and predicates that can be introduced in a recursive extension of PA are recursive.

Proposition 7.4.3 *Let R be an n-ary predicate on $\mathbb{N}$. Then R can be introduced in a recursive extension of PA if and only if χ_R can be introduced.*

Proof. Suppose R has been introduced in a recursive extension of PA with the defining axiom an open formula φ. Then we can introduce χ_R by

$$\chi_R(v_1, \ldots, v_n) = \mu w((\varphi(v_1, \ldots, v_n) \wedge w = 0) \vee (\neg\varphi(v_1, \ldots, v_n) \wedge w = 1)).$$

Now assume that χ_R has been introduced in a recursive extension P' of PA. Then the formula

$$\chi_R(v_1, \ldots, v_n) = 0$$

introdues R. $\square$

Example 7.4.4 The functions $+$ (addition) and $\cdot$ (multiplication) are nonlogical symbols of PA. By Example 7.4.1, each projection map π_i^n can be introduced. Since $<$ is a nonlogical symbol of PA, by Proposition 7.4.3, $\chi_<$ can be introduced. Thus, all initial recursive functions can be introduced in a recursive extension of PA.

Example 7.4.5 Let $n \geq 1$ and $p \in \mathbb{N}$. We can introduce the n-ary constant function C_p^n by taking the formula $\varphi[w, v_1, \cdots, v_n]$ to be $w = k_p$.

Example 7.4.6 We can introduce $\dot{-}$ to PA by

$$x \dot{-} y = \mu z(x + z = y \vee x < y)$$

by the above method.

Proposition 7.4.7 *The set of functions that can be introduced in a recursive extension of PA is closed under composition.*

Proof. Let $f_1, \ldots, f_k$ be n-ary functions and g a k-ary function that have been introduced in a recursive extension P' of PA. Further assume that

$$f_i v_1 \cdots v_n = \mu w_i \varphi_i[w_i, v_1, \ldots, v_n], \quad 1 \leq i \leq k,$$

and

$$g w_1 \cdots w_k = \mu w \varphi[w, w_1, \ldots, w_k].$$

Suppose

$$h(m_1, \ldots, m_n) = g(f_1(m_1, \ldots, m_n), \ldots, f_k(m_1, \ldots, m_n)).$$

It is not difficult to prove that

$$P' \vdash \exists w[w = g(f_1(v_1 \ldots, v_n), \ldots, f_k(v_1, \ldots, v_n))].$$

Hence, we can introduce h as follows:

$$h v_1 \cdots v_n = \mu w[w = g(f_1(v_1 \ldots, v_n), \ldots, f_k(v_1, \ldots, v_n))].$$

□

Remark 7.4.8 Let g be an $(n+1)$-ary function that has been introduced in a recursive extension P' of PA and let

$$\forall m_1 \cdots \forall m_n \exists l[g(m_1, \ldots, m_n, l) = 0].$$

Define

$$h(m_1, \ldots, m_n) = \mu l[g(m_1, \ldots, m_n, l) = 0].$$

Suppose

$$g v_1 \cdots v_{n+1} = \mu w \varphi[w, v_1, \ldots, v_{n+1}].$$

If

$$P \vdash \exists v_{n+1}[g v_1 \cdots v_n v_{n+1} = 0],$$

then we can introduce h as follows:

$$h v_1 \cdots v_n = \mu v_{n+1} g v_1 \cdots v_n v_{n+1} = 0.$$

A great many recursive functions and recursive predicates can be introduced in a recursive extension of PA. Further, the set of all functions and predicates that can be introduced in a recursive extension satisfies some of the closure properties satisfied by the set of recursive functions and recursive predicates. One can easily see this by looking at the explicit definitions of many recursive functions that we have defined earlier and the proofs of closure properties of the set of all recursive functions and recursive predicates.

Exercise 7.4.9 1. Let $f_1, \ldots, f_k$ be n-ary functions and P a k-ary predicate. Assume that $f_1, \ldots, f_k$ and P can be introduced. Show that the n-ary predicate Q defined by

$$Q(m_1, \ldots, m_n) \Leftrightarrow P(f_1(m_1, \ldots, m_n), \ldots, f_k(m_1, \ldots, m_n))$$

can be introduced.

2. Let P and Q be n-ary predicates that can be introduced. Show that the predicates $\neg P$, $P \vee Q$, $P \wedge Q$, $P \to Q$, and $P \leftrightarrow Q$ can be introduced.
3. Show that the set of all functions and predicates that can be introduced is closed under bounded minimalizations, and bounded quantifiers.

Exercise 7.4.10 Show that the divisibility $m|n$, the ordered pair function OP, and Gödel's β-function can be introduced in a recursive extension of PA.

Exercise 7.4.11 Let $A_1, \ldots, A_m$ be pairwise disjoint subsets of $\mathbb{N}^k$ whose union is $\mathbb{N}^k$. Suppose $f_1, \ldots, f_m$ are k-ary functions. Define $g : \mathbb{N}^k \to \mathbb{N}$ by

$$g(\overline{a}) = \begin{cases} f_1(\overline{a}) \text{ if } \overline{a} \in A_1, \\ \quad\vdots \\ f_m(\overline{a}) \text{ if } \overline{a} \in A_m. \end{cases}$$

Show that if $A_1, \ldots, A_m$ and $f_1, \ldots, f_m$ can be introduced in a recursive extension of PA, then g can be introduced.

Exercise 7.4.12 Show that the set of all functions that can be introduced in a recursive extension of PA is closed under primitive recursion.

Exercise 7.4.13 Show that all the finitely many functions and predicates for PA that were introduced in the section on arithmetization of theories can be introduced in a recursive extension of PA.

We now proceed to prove that every true closed existential formula of a recursive extension P' of PA is a theorem of P'.

In the sequel we shall use the same notation for the functions and predicates introduced. For instance, we shall use form *both for the predicate* form(n) *as well as for the corresponding function symbol introduced,* num *both for the function* num(n) *as well as for the corresponding function symbol introduced, and* Prf_{PA} *for the predicate* $\mathrm{Prf}_{PA}(m,n)$ *as well as for the corresponding symbol, and so on.*

Let P' be a recursive extension of PA. The set of **R-formulas** of P' is the smallest class of formulas $\mathcal{F}$ that contains all formulas of the form $fv_1 \cdots v_n = v$, $pv_1 \cdots v_n$, and $\neg pv_1 \cdots v_n$ (f and p function and predicate symbols of P') and that satisfies

(a) $A, B \in \mathcal{F} \Rightarrow A \vee B, A \wedge B \in \mathcal{F}$.
(b) If $A \in \mathcal{F}$ and if x, y are distinct variables, then $\forall x(x < y \rightarrow A) \in \mathcal{F}$.
(c) If $A \in \mathcal{F}$, $\exists x A \in \mathcal{F}$.

A formula of PA of the form $\varphi_{v_1,\ldots,v_m}[k_{n_1},\ldots,k_{n_m}]$, $n_1,\ldots,n_m \in \mathbb{N}$, will be called a **numerical instance** of φ.

Proposition 7.4.14 *Let A be an R-formula of PA. Then every true numerical instance of A is a theorem of PA.*

Proof. That R-formulas of the form $x = y$, $Sx = y$, $x+y = z$, $x \cdot y = z$, $x < y$ and negations of these formulas satisfy the conclusion of the proposition follows from the representability of $=$, S, $+$, $\cdot$, and $<$ and the fact that PA is an extension of N.

Now we show that the set $\mathcal{G}$ of formulas that satisfy the conclusion of the proposition satisfy the three closure properties (a)–(c). If A, B satisify the conclusion, it is quite easy to check that $A \vee B$ and $A \wedge B$ also satisfy the conclusion. Thus, (a) holds for $\mathcal{G}$.

We now show that (c) holds for $\mathcal{G}$. Let A satisfy the conclusion of the proposition and let B be the formula $\exists x A$. A numerical instance B' of B is of the form $\exists x A'$, where A' is obtained by substituting numerals in A for all free variables other than x. Suppose B' is true. Then $A'_x[k_n]$ is true for some n. By our assumption, $PA \vdash A'_x[k_n]$. This implies that $PA \vdash B'$ by the substitution axiom and the detachment rule.

Finally, we show that (b) holds for $\mathcal{G}$. Let A satisfy the conclusion of the proposition and let B be the formula $\forall x(x < y \rightarrow A)$. A numerical instance B' of B is of the form $\forall x(x < k_n \rightarrow A')$, where A' is obtained by substituting numerals in A for all free variables other than x and k_n for y. Suppose B' is true. Then for each $i < n$, $A'_x[k_i]$ is true. Hence, by our hypothesis they are theorems of PA. By Lemma 4.3.1, the detachment rule, and the $\forall$-introduction rule,

$$PA \vdash B'.$$

□

Recall that a formula is called existential if it is in prenex form and all the quantifiers in its prefix are $\exists$.

Proposition 7.4.15 *Let P' be a recursive extension of PA. Then every existential formula A of P' is equivalent in P' to an R-formula.*

Proof. In view of the defining condition (c) of R-formulas, it is sufficient to prove that every open formula A of P' is equivalent in P' to an R-formula.

Step 1: Let $t[x_1, \ldots, x_n]$ be a term of P', and A the formula $x = t$. The result holds for A.

We proceed by induction on the length of t. If t is a variable, then A is an R-formula. If t is a constant, then $t = x$ is an R-formula. This is equivalent to $x = t$ by the symmetry theorem. Now let $t = ft_1 \cdots t_n$. Then, by Proposition 4.2.23,

$$P' \vdash x = t \leftrightarrow \exists y_1 \cdots \exists y_n (y_1 = t_1 \wedge \cdots \wedge y_n = t_n \wedge x = fy_1 \cdots y_n).$$

The result now follows from the induction hypothesis, the symmetry theorem, and the definition of R-formulas.

Step 2: Let A be a formula of the form $pt_1 \cdots t_n$, p a relation symbol, and $t_1, \cdots, t_n$ terms. We have

$$P' \vdash A \leftrightarrow \exists y_1 \cdots \exists y_n (y_1 = t_1 \wedge \cdots \wedge y_n = t_n \wedge py_1 \cdots y_n).$$

Since each formula $y_i = t_i$ is equivalent to an R-formula, the result holds for A.

Similarly, we prove the result for formulas that are negations of atomic formulas.

The class of formulas for which the theorem holds is, by the defining condition (a) of R-formulas, closed under $\vee$ and $\wedge$. Hence, the result for open formulas follows from Exercise 4.2.24. □

Proposition 7.4.16 *Let P' be a recursive extension of PA. Then every R-formula of P' is equivalent in P' to an R-formula in PA.*

Proof. Suppose P'' is a recursive extension of P' obtained by adding just one nonlogical symbol. Our result will be proved if we show that every R-formula A in P'' is equivalent in P'' to an R-formula in P'. By the equivalence theorem, it is sufficient to prove the result for A of the form $fx_1 \cdots x_n = y$ or $px_1 \cdots x_n$ or their negations.

Since the defining axiom of p is an open formula of P', the result is easy to prove for a formula of the form $px_1 \cdots x_n$ and their negations by Proposition 7.4.15. Since a formula of the form $\neg(fx_1 \cdots x_n = y)$ is equivalent in P'' to a formula $\exists z(\neg(y = z) \wedge fx_1 \cdots x_n = z)$, the result follows for such formulas by Proposition 7.4.15.

Let A be $fx_1 \cdots x_n = y$. Then A is equivalent to a formula of P' of the form

$$B \wedge \forall x(x < y \rightarrow C),$$

where B and C are open. The result can be easily seen now by Proposition 7.4.15. □

From the last three results, we have the following theorem.

Theorem 7.4.17 *If P' is a recursive extension of PA, then every true closed existential formula is a theorem of P'.*

7.5 Second Incompleteness Theorem

We are now in a position to prove that "PA is consistent" is not a theorem of PA.

From now on, we assume that P' is a recursive extension of PA in which all the recursive functions and recursive predicates for PA introduced in the section on arithmetization of theories have been introduced. We introduce some notation first.

Let t, $t_1, \ldots, t_n$ be terms and $x_1, \ldots, x_n$ be the first n variables in alphabetical order. We define the terms $S(t, t_1, \ldots, t_n)$ by induction:

$$S(t, t_1) = \text{sub}(t, k_{\ulcorner x_1 \urcorner}, t_1),$$

$$S(t, t_1, t_2) = \text{sub}(S(t, t_1), k_{\ulcorner x_2 \urcorner}, t_2)$$

$$\vdots$$

$$S(t, t_1, \ldots, t_n) = \text{sub}(S(t, t_1, \ldots, t_{n-1}), k_{\ulcorner x_n \urcorner}, t_n).$$

Since each formula in P' has a translation in PA and since P' is a conservative extension of PA, we shall not distinguish between a formula of P' and its translation in P. With these conventions, we abbreviate the formula $\exists y \text{Prf}_{PA}(x, y)$ by $\text{Thm}_{PA}(x)$ and

$$\neg \forall x (\text{form}_{PA}(x) \to \text{Thm}_{PA}(x))$$

by Con_{PA}.

By formalizing the proof of Proposition 7.4.14 inside P', we shall prove the following lemma.

Lemma 7.5.1 *For any R-formula $A[x_1, \ldots, x_n]$ of PA,*

$$P' \vdash A \to \text{Thm}_{PA}(S(k_{\ulcorner A \urcorner}, x_1, \ldots, x_n)).$$

Proof. We shall prove the result by induction on the length of A. We shall give only a few steps of the proof. Readers should not find it difficult to complete the proof themselves.

Let A be the formula $0 = x$. We have

$$S(k_{\ulcorner A \urcorner}, x) = \langle k_{SN(=)}, \text{num}(0), \text{num}(x) \rangle.$$

We have to show that

$$P' \vdash 0 = x \to \mathrm{Thm}_{PA}(\langle k_{SN(=)}, \mathrm{num}(0), \mathrm{num}(0)\rangle).$$

By the equality theorem, this will be proved if we show that

$$P' \vdash \mathrm{Thm}_{PA}(\langle k_{SN(=)}, \mathrm{num}(0), \mathrm{num}(0)\rangle).$$

But the formula

$$\mathrm{Thm}_{PA}(\langle k_{SN(=)}, \mathrm{num}(0), \mathrm{num}(0)\rangle)$$

is a true closed existential formula. Hence it is a theorem of P' by Theorem 7.4.17.

Similarly by formalizing the proofs of the representability of S, $+$, $\cdot$, and $<$ inside P', we can prove the assertion for formulas of the form $Sx = y$, $x + y = z$, $x \cdot y = z$, $x < y$, etc.

We shall show only one inductive step and leave the others for the reader to prove. Let $A[x_1, \ldots, x_n]$ be of the form $\exists xB$ and the result holds for the formula B. Set

$$t = S(k_{\lceil B \rceil}, x_1, \ldots, x_n).$$

Since $S(k_{\lceil A \rceil}, x_1, \ldots, x_n) = \langle k_{SN(\exists)}, k_{\lceil x \rceil}, t\rangle$ is a true closed existential formula, by Theorem 7.4.17,

$$P' \vdash S(k_{\lceil A \rceil}, x_1, \ldots, x_n) = \langle k_{SN(\exists)}, k_{\lceil x \rceil}, t\rangle. \qquad (a)$$

By the induction hypothesis,

$$P' \vdash B \to \mathrm{Thm}_{PA}(\mathrm{sub}(t, k_{\lceil x \rceil}, \mathrm{num}(x))).$$

By the distribution rule, we have

$$P' \vdash A \to \exists x \mathrm{Thm}_{PA}(\mathrm{sub}(t, k_{\lceil x \rceil}, \mathrm{num}(x))). \qquad (b)$$

Using the fact that every true existential sentence is a theorem, we can formalize the proof of "if $P' \vdash B_x[k_n]$ for some n, $P' \vdash \exists xB$" inside P'. Thus, we get

$$P' \vdash \exists x \mathrm{Thm}_{PA}(\mathrm{sub}(t, k_{\lceil x \rceil}, \mathrm{num}(x))) \to \mathrm{Thm}_{P}(\langle k_{SN(\exists)}, k_{\lceil x \rceil}, t\rangle). \qquad (c)$$

Our assertion for A follows from (a), (b), and (c). □

Essentially by formalizing the proof of the first incompleteness theorem inside P', we get the following very interesting result.

Theorem 7.5.2 (Second incompleteness theorem) *Con_{PA} is not a theorem of PA.*

Proof. Let $A[x]$ denote the translation of the formula

$$\neg\exists y \mathrm{Prf}_{PA}(sub(x, k_{\ulcorner x \urcorner}, \mathrm{num}(x)), y)$$

in PA. Let

$$a = \ulcorner A \urcorner .$$

We need to show that

$$PA \nvdash \mathrm{Con}_{PA}. \tag{1}$$

Since P' is an extension by definition of PA, it is sufficient to prove that

$$P' \nvdash \mathrm{Con}_{PA}. \tag{2}$$

For this, it is enough to prove that

$$P' \vdash \mathrm{Con}_{PA} \rightarrow A_x[k_a], \tag{3}$$

because by the argument contained in the proof of the first incompleteness theorem,

$$P' \nvdash A_x[k_a]$$

and $A_x[k_a]$ is a tautological consequence of $\mathrm{Con}_{PA} \rightarrow A_x[k_a]$ and Con_{PA}.

Let B be an R-formula in PA that is equivalent in P' to

$$\exists y \mathrm{Prf}_{PA}(\mathrm{sub}(x, k_{\ulcorner x \urcorner}, \mathrm{num}(x))), y)_x[k_a].$$

Hence, by the equivalence theorem,

$$P' \vdash \neg B \leftrightarrow A_x[k_a]. \tag{4}$$

Let $b = \ulcorner B \urcorner$ and $c = \ulcorner A_x[k_a] \urcorner$. By the definition of consistency and Theorem 7.4.17, we have

$$P' \vdash \mathrm{Con}_{PA} \rightarrow (\neg \mathrm{Thm}_{PA}(k_c) \vee \neg \mathrm{Thm}_{PA}(neg(k_c))), \tag{5}$$

where

$$neg(k_c) = \langle k_{SN(\neg)}, c \rangle.$$

Thus, by the tautology theorem, it is sufficient to show that

$$P' \vdash (\neg \mathrm{Thm}_{PA}(k_c) \vee \neg \mathrm{Thm}_{PA}(neg(k_c))) \rightarrow A_x[k_a]. \tag{6}$$

By the tautology theorem again, it is enough to show that

$$P' \vdash \neg \mathrm{Thm}_{PA}(k_c) \rightarrow A_x[k_a] \tag{7}$$

and

$$P' \vdash \neg \mathrm{Thm}_{PA}(neg(k_c)) \rightarrow A_x[k_a]. \tag{8}$$

We prove (7) as follows:

Note that

$$k_c = \text{sub}(k_a, k_{\lceil x \rceil}, \text{num}(k_a))$$

is a true closed existential formula of P'. Hence, by Theorem 7.4.17, it is a theorem of P'. By the equality theorem and the definition of A, we now have

$$P' \vdash \neg A_x[k_a] \to \text{Thm}_{PA}(k_c).$$

This proves (7).

To prove (8), first note that by (4),

$$P' \vdash B \to \neg A_x[k_a].$$

Hence,

$$\text{Thm}_{PA}(\langle k_{SN(\vee)}, neg(k_b), neg(k_c)\rangle)$$

is a true closed existential formula. Hence, by Theorem 7.4.17,

$$P' \vdash \text{Thm}_{PA}(k_b) \to \text{Thm}_{PA}(neg(k_c)), \tag{9}$$

Finally, by Lemma 7.5.1, we have,

$$P' \vdash B \to \text{Thm}_{PA}(k_b). \tag{10}$$

Now note that

$$\neg\text{Thm}_{PA}(neg(k_c)) \to A_x[k_a]$$

is a tautological consequence of

$$\text{Thm}_{PA}(k_b) \to \text{Thm}_{PA}(neg(k_c)),$$

$$B \to \text{Thm}_{PA}(k_b),$$

and

$$\neg B \leftrightarrow A_x[k_a].$$

Hence, (8) follows from (9), (10), and (4) by the tautology theorem. □

Remark 7.5.3 There is an extension by definitions of ZF (or of ZFC) in which there is a suitable interpretation of Peano arithmetic PA so that the representability theorem can be proved with the theory N replaced by ZF. Hence, we can express Con_{ZF} and Con_{ZFC} as formulas of ZF. Again, using similar ideas, we can prove the following result:

$$ZF \nvdash Con_{ZF}.$$

References

[1] J. Ax. The elementary theory of finite fields. Annals of Math. 88 (1968), 103–115.

[2] C. C. Chang and H. J. Keisler. *Model Theory.* North-Holland Publishing Company, 1990.

[3] P. Hinman. *Fundamentals of Mathematical Logic.* A. K. Peters Ltd., 2005.

[4] D. R. Hofstadter. *Gödel, ESCHER, BACH: An Eternal Golden Braid.* Vintage Books, 1989.

[5] T. Jech. *Set Theory.* Springer Monographs in Mathematics, 2006.

[6] K. Kunnen. *Set Theory: An Introduction to Independence Proofs.* North-Holland Publishing Company, 1980

[7] S. Lang. *Algebra.* Addison-Wesley, 1971

[8] D. Marker. *Model Theory: An Introduction.* GTM 217, Springer, 2002.

[9] H. J. Rogers. *Theory of Recursive Functions and Effective Computability.* McGraw-Hill Book Company, New York, 1967.

[10] R. Penrose. *The Emperor's New Mind.* Oxford University Press, Oxford, New York, Melbourne, 1990.

[11] J. R. Shoenfield. *Mathematical Logic.* A. K. Peters Ltd., 2001.

[12] S. M. Srivastava. *A Course on Borel Sets.* GTM 180, Springer, 1998.

Index

Universitext

Aguilar, M.; Gitler, S.; Prieto, C.: Algebraic Topology from a Homotopical Viewpoint

Aksoy, A.; Khamsi, M. A.: Methods in Fixed Point Theory

Alevras, D.; Padberg M. W.: Linear Optimization and Extensions

Andersson, M.: Topics in Complex Analysis

Aoki, M.: State Space Modeling of Time Series

Arnold, V. I.: Lectures on Partial Differential Equations

Arnold, V. I.; Cooke, R.: Ordinary Differential Equations

Audin, M.: Geometry

Aupetit, B.: A Primer on Spectral Theory

Bachem, A.; Kern, W.: Linear Programming Duality

Bachmann, G.; Narici, L.; Beckenstein, E.: Fourier and Wavelet Analysis

Badescu, L.: Algebraic Surfaces

Balakrishnan, R.; Ranganathan, K.: A Textbook of Graph Theory

Balser, W.: Formal Power Series and Linear Systems of Meromorphic Ordinary Differential Equations

Bapat, R.B.: Linear Algebra and Linear Models

Benedetti, R.; Petronio, C.: Lectures on Hyperbolic Geometry

Benth, F. E.: Option Theory with Stochastic Analysis

Berberian, S. K.: Fundamentals of Real Analysis

Berger, M.: Geometry I, and II

Bhattacharya, R.; Waymire, E.C.: A Basic Course in Probability Theory

Bliedtner, J.; Hansen, W.: Potential Theory

Blowey, J. F.; Coleman, J. P.; Craig, A. W. (Eds.): Theory and Numerics of Differential Equations

Blowey, J. F.; Craig, A.; Shardlow, T. (Eds.): Frontiers in Numerical Analysis, Durham 2002, and Durham 2004

Blyth, T. S.: Lattices and Ordered Algebraic Structures

Börger, E.; Grädel, E.; Gurevich, Y.: The Classical Decision Problem

Böttcher, A; Silbermann, B.: Introduction to Large Truncated Toeplitz Matrices

Boltyanski, V.; Martini, H.; Soltan, P. S.: Excursions into Combinatorial Geometry

Boltyanskii, V. G.; Efremovich, V. A.: Intuitive Combinatorial Topology

Bonnans, J. F.; Gilbert, J. C.; Lemarchal, C.; Sagastizbal, C. A.: Numerical Optimization

Booss, B.; Bleecker, D. D.: Topology and Analysis

Borkar, V. S.: Probability Theory

Brunt B. van: The Calculus of Variations

Bühlmann, H.; Gisler, A.: A Course in Credibility Theory and its Applications

Carleson, L.; Gamelin, T. W.: Complex Dynamics

Cecil, T. E.: Lie Sphere Geometry: With Applications of Submanifolds. 2nd edition

Chae, S. B.: Lebesgue Integration

Chandrasekharan, K.: Classical Fourier Transform

Charlap, L. S.: Bieberbach Groups and Flat Manifolds

Chern, S.: Complex Manifolds without Potential Theory

Chorin, A. J.; Marsden, J. E.: Mathematical Introduction to Fluid Mechanics

Cohn, H.: A Classical Invitation to Algebraic Numbers and Class Fields

Curtis, M. L.: Abstract Linear Algebra

Curtis, M. L.: Matrix Groups

Cyganowski, S.; Kloeden, P.; Ombach, J.: From Elementary Probability to Stochastic Differential Equations with MAPLE

Da Prato, G.: An Introduction to Infinite Dimensional Analysis

Dalen, D. van: Logic and Structure

Das, A.: The Special Theory of Relativity: A Mathematical Exposition

Debarre, O.: Higher-Dimensional Algebraic Geometry

Deitmar, A.: A First Course in Harmonic Analysis

Demazure, M.: Bifurcations and Catastrophes

Devlin, K. J.: Fundamentals of Contemporary Set Theory

DiBenedetto, E.: Degenerate Parabolic Equations

Diener, F.; Diener, M.(Eds.): Nonstandard Analysis in Practice

Dimca, A.: Sheaves in Topology

Dimca, A.: Singularities and Topology of Hypersurfaces

DoCarmo, M. P.: Differential Forms and Applications

Duistermaat, J. J.; Kolk, J. A. C.: Lie Groups

Dumortier.: Qualitative Theory of Planar Differential Systems

Dundas, B. I.; Levine, M.; Østvaer, P. A.; Röndip, O.; Voevodsky, V.: Motivic Homotopy Theory

Edwards, R. E.: A Formal Background to Higher Mathematics Ia, and Ib

Edwards, R. E.: A Formal Background to Higher Mathematics IIa, and IIb

Emery, M.: Stochastic Calculus in Manifolds

Emmanouil, I.: Idempotent Matrices over Complex Group Algebras

Endler, O.: Valuation Theory

Engel, K.-J.; Nagel, R.: A Short Course on Operator Semigroups

Erez, B.: Galois Modules in Arithmetic

Everest, G.; Ward, T.: Heights of Polynomials and Entropy in Algebraic Dynamics

Farenick, D. R.: Algebras of Linear Transformations

Foulds, L. R.: Graph Theory Applications

Franke, J.; Hrdle, W.; Hafner, C. M.: Statistics of Financial Markets: An Introduction

Frauenthal, J. C.: Mathematical Modeling in Epidemiology

Freitag, E.; Busam, R.: Complex Analysis

Friedman, R.: Algebraic Surfaces and Holomorphic Vector Bundles

Fuks, D. B.; Rokhlin, V. A.: Beginner's Course in Topology

Fuhrmann, P. A.: A Polynomial Approach to Linear Algebra

Gallot, S.; Hulin, D.; Lafontaine, J.: Riemannian Geometry

Gardiner, C. F.: A First Course in Group Theory

Gårding, L.; Tambour, T.: Algebra for Computer Science

Godbillon, C.: Dynamical Systems on Surfaces

Godement, R.: Analysis I, and II

Goldblatt, R.: Orthogonality and Spacetime Geometry

Gouvêa, F. Q.: p-Adic Numbers

Gross, M. et al.: Calabi-Yau Manifolds and Related Geometries

Gustafson, K. E.; Rao, D. K. M.: Numerical Range. The Field of Values of Linear Operators and Matrices

Gustafson, S. J.; Sigal, I. M.: Mathematical Concepts of Quantum Mechanics

Hahn, A. J.: Quadratic Algebras, Clifford Algebras, and Arithmetic Witt Groups

Hájek, P.; Havránek, T.: Mechanizing Hypothesis Formation

Heinonen, J.: Lectures on Analysis on Metric Spaces

Hlawka, E.; Schoißengeier, J.; Taschner, R.: Geometric and Analytic Number Theory

Holmgren, R. A.: A First Course in Discrete Dynamical Systems

Howe, R., Tan, E. Ch.: Non-Abelian Harmonic Analysis

Howes, N. R.: Modern Analysis and Topology

Hsieh, P.-F.; Sibuya, Y. (Eds.): Basic Theory of Ordinary Differential Equations

Humi, M., Miller, W.: Second Course in Ordinary Differential Equations for Scientists and Engineers

Hurwitz, A.; Kritikos, N.: Lectures on Number Theory

Huybrechts, D.: Complex Geometry: An Introduction

Isaev, A.: Introduction to Mathematical Methods in Bioinformatics

Istas, J.: Mathematical Modeling for the Life Sciences

Iversen, B.: Cohomology of Sheaves

Jacod, J.; Protter, P.: Probability Essentials

Jennings, G. A.: Modern Geometry with Applications

Jones, A.; Morris, S. A.; Pearson, K. R.: Abstract Algebra and Famous Inpossibilities

Jost, J.: Compact Riemann Surfaces

Jost, J.: Dynamical Systems. Examples of Complex Behaviour

Jost, J.: Postmodern Analysis

Jost, J.: Riemannian Geometry and Geometric Analysis

Kac, V.; Cheung, P.: Quantum Calculus

Kannan, R.; Krueger, C. K.: Advanced Analysis on the Real Line

Kelly, P.; Matthews, G.: The Non-Euclidean Hyperbolic Plane

Kempf, G.: Complex Abelian Varieties and Theta Functions

Kitchens, B. P.: Symbolic Dynamics

Klenke, A.: Probability Theory: A Comprehensive Course

Kloeden, P.; Ombach, J.; Cyganowski, S.: From Elementary Probability to Stochastic Differential Equations with MAPLE

Kloeden, P. E.; Platen; E.; Schurz, H.: Numerical Solution of SDE Through Computer Experiments

Koralov, L.; Sina, Ya. G.: Theory of Probability and Random Processes. 2nd edition

Kostrikin, A. I.: Introduction to Algebra

Krasnoselskii, M. A.; Pokrovskii, A. V.: Systems with Hysteresis

Kuo, H.-H.: Introduction to Stochastic Integration

Kurzweil, H.; Stellmacher, B.: The Theory of Finite Groups. An Introduction

Kyprianou, A.E.: Introductory Lectures on Fluctuations of Lévy Processes with Applications

Lang, S.: Introduction to Differentiable Manifolds

Lefebvre, M.: Applied Stochastic Processes

Lorenz, F.: Algebra I: Fields and Galois Theory

Lorenz, F.: Algebra II: Fields with Structure, Algebras and Advanced Topics

Luecking, D. H., Rubel, L. A.: Complex Analysis. A Functional Analysis Approach

Ma, Zhi-Ming; Roeckner, M.: Introduction to the Theory of (non-symmetric) Dirichlet Forms

Mac Lane, S.; Moerdijk, I.: Sheaves in Geometry and Logic

Marcus, D. A.: Number Fields

Martinez, A.: An Introduction to Semiclassical and Microlocal Analysis

Matoušek, J.: Using the Borsuk-Ulam Theorem

Matsuki, K.: Introduction to the Mori Program

Mazzola, G.; Milmeister G.; Weissman J.: Comprehensive Mathematics for Computer Scientists 1

Mazzola, G.; Milmeister G.; Weissman J.: Comprehensive Mathematics for Computer Scientists 2

Mc Carthy, P. J.: Introduction to Arithmetical Functions

McCrimmon, K.: A Taste of Jordan Algebras

Meyer, R. M.: Essential Mathematics for Applied Field

Meyer-Nieberg, P.: Banach Lattices

Mikosch, T.: Non-Life Insurance Mathematics

Mines, R.; Richman, F.; Ruitenburg, W.: A Course in Constructive Algebra

Moise, E. E.: Introductory Problem Courses in Analysis and Topology

Montesinos-Amilibia, J. M.: Classical Tessellations and Three Manifolds

Morris, P.: Introduction to Game Theory

Mortveit, H.; Reidys, C.: An Introduction to Sequential Dynamical Systems

Nicolaescu, L.: An Invitation to Morse Theory

Nikulin, V. V.; Shafarevich, I. R.: Geometries and Groups

Oden, J. J.; Reddy, J. N.: Variational Methods in Theoretical Mechanics

Øksendal, B.: Stochastic Differential Equations. 6th edition

Øksendal, B.; Sulem, A.: Applied Stochastic Control of Jump Diffusions. 2nd edition

Orlik, P.; Welker, V.: Algebraic Combinatorics

Perrin, D.: Algebraic Geometry: An Introduction

Poizat, B.: A Course in Model Theory

Polster, B.: A Geometrical Picture Book

Porter, J. R.; Woods, R. G.: Extensions and Absolutes of Hausdorff Spaces

Procesi, C.: Lie Groups

Radjavi, H.; Rosenthal, P.: Simultaneous Triangularization

Ramsay, A.; Richtmeyer, R. D.: Introduction to Hyperbolic Geometry

Rautenberg, W.: A concise Introduction to Mathematical Logic

Rees, E. G.: Notes on Geometry

Reisel, R. B.: Elementary Theory of Metric Spaces

Rey, W. J. J.: Introduction to Robust and Quasi-Robust Statistical Methods

Ribenboim, P.: Classical Theory of Algebraic Numbers

Rickart, C. E.: Natural Function Algebras

Rotman, J. J.: Galois Theory

Rubel, L. A.: Entire and Meromorphic Functions

Ruiz-Tolosa, J. R.; Castillo E.: From Vectors to Tensors

Runde, V.: A Taste of Topology

Rybakowski, K. P.: The Homotopy Index and Partial Differential Equations

Sabbah, C.: Isomonodromic Deformations and Frobenius Manifolds: An introduction

Sagan, H.: Space-Filling Curves

Samelson, H.: Notes on Lie Algebras

Sauvigny, F.: Partial Differential Equations I

Sauvigny, F.: Partial Differential Equations II

Schiff, J. L.: Normal Families

Schirotzek, W.: Nonsmooth Analysis

Sengupta, J. K.: Optimal Decisions under Uncertainty

Séroul, R.: Programming for Mathematicians

Seydel, R.: Tools for Computational Finance

Shafarevich, I. R.: Discourses on Algebra

Shapiro, J. H.: Composition Operators and Classical Function Theory

Simonnet, M.: Measures and Probabilities

Smith, K. E.; Kahanpää, L.; Kekäläinen, P.; Traves, W.: An Invitation to Algebraic Geometry

Smith, K. T.: Power Series from a Computational Point of View

Smoryński, C.: Logical Number Theory I. An Introduction

Srivastava, S.M.: A Course on Mathematical Logic

Stichtenoth, H.: Algebraic Function Fields and Codes

Stillwell, J.: Geometry of Surfaces

Stroock, D. W.: An Introduction to the Theory of Large Deviations

Sunder, V. S.: An Invitation to von Neumann Algebras

Tamme, G.: Introduction to Étale Cohomology

Tondeur, P.: Foliations on Riemannian Manifolds

Toth, G.: Finite Mbius Groups, Minimal Immersions of Spheres, and Moduli

Tu, L. W.: An Introduction to Manifolds

Verhulst, F.: Nonlinear Differential Equations and Dynamical Systems

Weintraub, S. H.: Galois Theory

Wong, M. W.: Weyl Transforms

Xambó-Descamps, S.: Block Error-Correcting Codes

Zaanen, A.C.: Continuity, Integration and Fourier Theory

Zhang, F.: Matrix Theory

Zong, C.: Sphere Packings

Zong, C.: Strange Phenomena in Convex and Discrete Geometry

Zorich, V. A.: Mathematical Analysis I

Zorich, V. A.: Mathematical Analysis II

Made in the USA
Lexington, KY
08 December 2010